SCALE the UNIVERSE

What if I were so large (10^8 times my actual size) that I could stride across the solar system?

SCIENCE WITH SIMPLE THINGS SERIES

Conceived and written by **RON MARSON**

Illustrated by **PEG MARSON**

TOPS LEARNING SYSTEMS

342 S Plumas Street
Willows, CA 95988

www.topscience.org

WHAT CAN YOU COPY?

Dear Educator,

Please honor our copyright restrictions. We offer liberal options and guidelines below with the intention of balancing your needs with ours. When you buy these labs and use them for your own teaching, you sustain our work. If you "loan" or circulate copies to others without compensating TOPS, you squeeze us financially, and make it harder for our small non-profit to survive. Our well-being rests in your hands. Please help us keep our low-cost, creative lessons available to students everywhere. Thank you!

PURCHASE, ROYALTY and LICENSE OPTIONS

TEACHERS, HOMESCHOOLERS, LIBRARIES:

We do all we can to keep our prices low. Like any business, we have ongoing expenses to meet. We trust our users to observe the terms of our copyright restrictions. While we prefer that all users purchase their own TOPS labs, we accept that real-life situations sometimes call for flexibility.

Reselling, trading, or loaning our materials is prohibited unless one or both parties contribute an Honor System Royalty as fair compensation for value received. We suggest the following amounts – let your conscience be your guide.

HONOR SYSTEM ROYALTIES: If making copies from a library, or sharing copies with colleagues, please calculate their value at 50 cents per lesson, or 25 cents for homeschoolers. This contribution may be made at our website or by mail (addresses at the bottom of this page). Any additional tax-deductible contributions to make our ongoing work possible will be accepted gratefully and used well.

Please follow through promptly on your good intentions. Stay legal, and do the right thing.

SCHOOLS, DISTRICTS, and HOMESCHOOL CO-OPS:

PURCHASE Option: Order a book in quantities equal to the number of target classrooms or homes, and receive quantity discounts. If you order 5 books or downloads, for example, then you have unrestricted use of this curriculum for any 5 classrooms or families per year for the life of your institution or co-op.

2-9 copies of any title: 90% of current catalog price + shipping.

10+ copies of any title: 80% of current catalog price + shipping.

ROYALTY/LICENSE Option: Purchase just one book or download *plus* photocopy or printing rights for a designated number of classrooms or families. If you pay for 5 additional Licenses, for example, then you have purchased reproduction rights for an entire book or download edition for any **6** classrooms or families per year for the life of your institution or co-op.

1-9 Licenses: 70% of current catalog price per designated classroom or home.

10+ Licenses: 60% of current catalog price per designated classroom or home.

WORKSHOPS and TEACHER TRAINING PROGRAMS:

We are grateful to all of you who spread the word about TOPS. Please limit copies to only those lessons you will be using, and collect all copyrighted materials afterward. No take-home copies, please. Copies of copies are strictly prohibited.

Copyright © 2005 by TOPS Learning Systems. All rights reserved. This material is created/printed/transmitted in the United States of America. No part of this program may be used, reproduced, or transmitted in any manner whatsoever without written permission from the publisher, *except as explicitly stated above and below*:

The *original owner* of this book or digital download is permitted to make multiple copies of all *student materials* for personal teaching use, provided all reproductions bear copyright notice. A purchasing school or homeschool co-op may assign *one* purchased book or digital download to *one* teacher, classroom, family, or study group *per year*. Reproduction of student materials from libraries is permitted if the user compensates TOPS as outlined above. Reproduction of any copyrighted materials for commercial sale is prohibited.

For licensing, honor system royalty payments, contact: **www.TOPScience.org**; or **TOPS Learning Systems 342 S Plumas St, Willows CA 95988**; or inquire at **customerservice@topscience.org**

ISBN 978-0-941008-44-0

This book is organized in **3 stand-alone sections**: you can teach them independently or all together. **Series A** reviews the base-ten logic of our metric system, and helps students comprehend exponential change by means of problem solving and unit analysis. In **Series B** students compare distances relative to each other, plot these on a log scale, and compute orders of magnitude. In **Series C**, students construct Books of Scale that tie the space-time universe together. The wide-ranging scale drawings and creative problem-solving challenges in this learning system result in an assessment portfolio students will be proud of.

Table of Contents

	PAGE
welcome	4
standards	4
why teach TOPS	5
photocopy table	6
getting ready	7
scope and sequence	8 – 9

A — Metric Longtape

	PAGE
A1: Orders of Magnitude	10 – 11
longtape	32
A2: Unit Analysis	12 – 13

B — Log Scales

	PAGE
B1: Ordering Distance	14 – 15
DISTANCE tabs (sticky version)	33
DISTANCE tabs (cutout version)	34
log scale	35 – 36
log scale KEY	37 – 38
B2: Using a Log Scale	16 – 17

C — Book of Scale

	PAGE
C1: Scale the Universe (1)	18 – 19
book of scale (line masters)	39 – 62
C2: Scale the Universe (2)	20 – 21
C3: Scale the Universe (3)	22 – 23
C4: Proportional Thinking	24 – 25
human scale ruler	64
C5: Ordering Time	26 – 31
TIME tabs (sticky version)	63
TIME tabs (cutout version)	64

Welcome, Dear Educator, ...

... to **SCALE THE UNIVERSE!** This is the second in a series of three books designed to inform students about NASA's 2007 GLAST space shot, while teaching standards-based math and science in engaging ways.

GLAST is an acronym that means **G**amma-ray **L**arge **A**rea **S**pace **T**elescope.* It will absorb, track and measure gamma ray photons across enormous energy ranges. Looking into the depths of solar flares, pulsars, quasars, active galactic nuclei, black holes, and gamma ray bursts, GLAST promises to help astronomers better understand interactions of matter at Nature's most extreme and energetic levels.

Our first GLAST book, FAR OUT MATH, examines how logarithms make sense of huge ranges of astronomical data. Students fold paper into slide rules, and use them for serious computing. Second-year algebra students and the mathematically curious will gain a concrete understanding of logarithms and how they work – adding and subtracting exponent distances to multiply and divide corresponding numbers.

In this second book, SCALE THE UNIVERSE, we explore logarithms further, within a context of orders of magnitude, metric measure, and scale drawing. We engage both high school and middle school students across 6 grade levels. How is this possible?

Compare a good hands-on science curriculum to a sandbox. Provide orientation, tools, and an interesting challenge, and kids young and old will learn something appropriate to their level of inquiry.

In SCALE THE UNIVERSE, we give you integrated activity clusters and astonishing flexibility: skip over some lessons, touch lightly on others, explore deeply where it suits your needs. Our inquiry-based investigations are enriched with elements of art, writing, imagination, and natural history on a grand scale, yet we minimize "cookbook" instruction. (See pages 8-9 for scope and sequence, subject and grade level recommendations).

Still to come: our final book in the NASA/GLAST trilogy will provide wide-ranging learning experiences with degrees, pi-radians, subtended angles, proportion, apparent size, parallax, latitude of Polaris, and other astronomical concepts.

Let us know about your teaching experiences with these books. Please direct your comments and suggestions to TOPS at the address on our copyright page, and send a copy to Lynn Cominsky, our program director.

Happy sciencing,
Ron Marson

*NOTE: GLAST has been renamed Fermi Gamma-ray Space Telescope.

SCALE THE UNIVERSE ties in with *National Math and Science Standards*.
See pages 8-9 for content details.

Science

The natural and designed world is complex; it is too large and complicated to investigate and comprehend all at once....

At the upper grades, the standard should facilitate and enhance the learning of scientific concepts and principles by providing students with a big picture of scientific ideas; for example, how measurement is important in all scientific endeavors.... An important part of measurement is knowing when to use which system....

Physical systems can be described at different levels of organization.... Further, systems at different levels of organization can manifest different properties and functions....

Scale includes understanding that different characteristics, properties, or relationships within a system might change as its dimensions are increased or decreased....

Thinking and analyzing in terms of systems will help students keep track of mass, energy, objects, organisms, and events referred to in the other content standards....

(The National Science Education Standards)

Math

Instructional programs from prekindergarten through grade 12 should enable all students to:

understand numbers, ways of representing numbers, relationships among numbers, and number systems;   understand measurable attributes of objects and the units, systems, and processes of measurement;   understand both metric and customary systems of measurement;   understand relationships among units and convert from one unit to another within the same system;   solve problems involving scale factors, using ratio and proportion;   make decisions about units and scales that are appropriate for problem situations involving measurement;   use unit analysis to check measurement computations;   create and use representations to organize, record, and communicate mathematical ideas;   use representations to model and interpret physical, social, and mathematical phenomena;   experience the rich interplay among mathematical topics, between mathematics and other subjects ...

(National Council of Teachers of Mathematics)

Why Teach TOPS?

Because your students have three brains – and multiple learning styles...

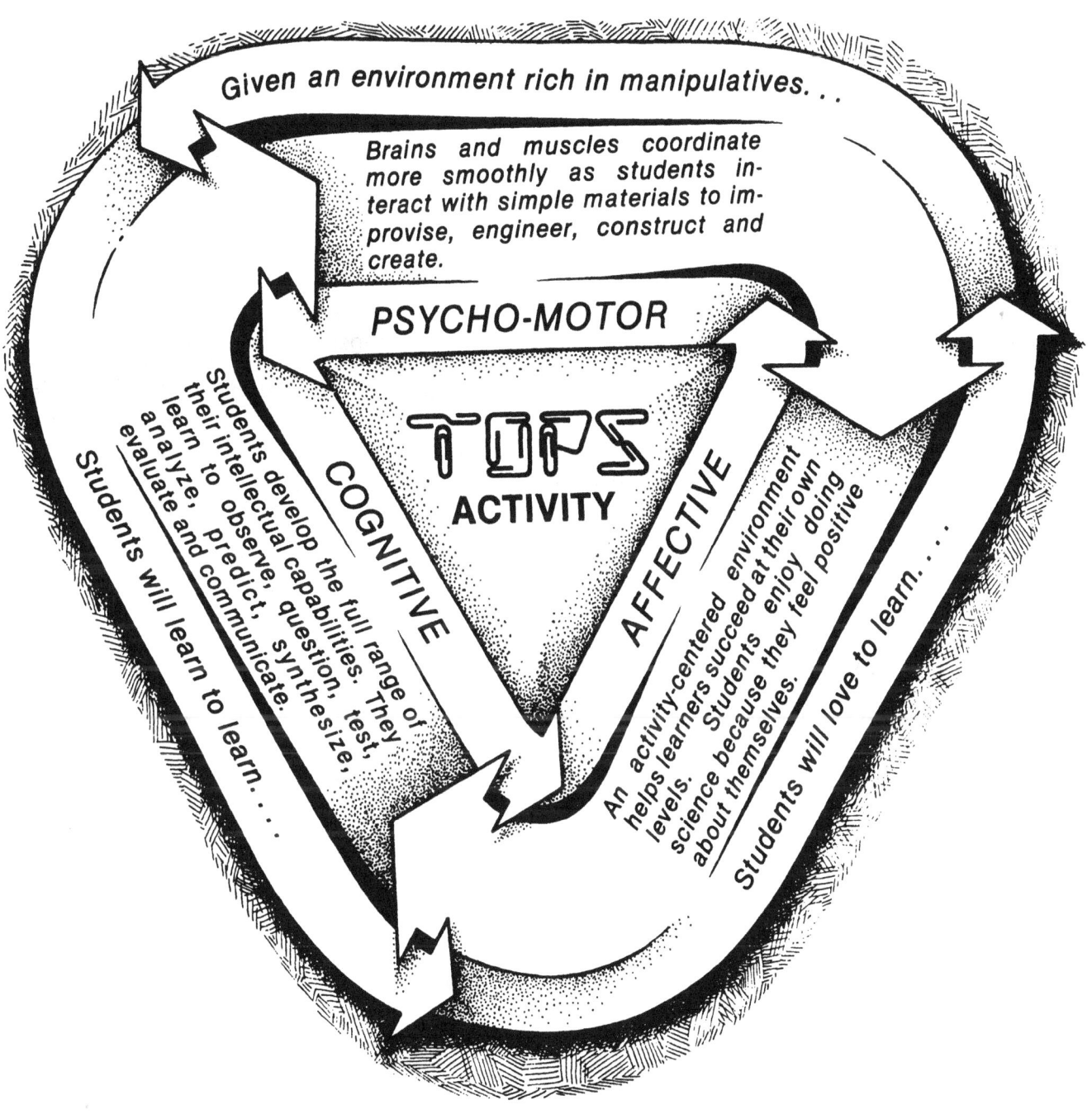

PHOTOCOPY TABLE

page title:	activity:	on page:	sides?	copies?	consumable?
Activity Sheets (9 pages)	A's, B's, C's	10–26	single (or double)	a class set	no: collate, staple
LongTape	A1/A2/B1/B2	32	single	1 per student	yes
DISTANCE Tabs (sticky)	B1	33	single	1 per class	yes
DISTANCE Tabs (cutout)	B1	34	single	1 per student	yes
Log Scale (worksheets)	B1/B2/C2	35–36	double (or single)	1 per student	yes
Log Scale (answer key)	B2/C2 (optional)	37–38	double (or single)	1 (or more)	no
Book of Scale	C1 through C5	39–62	double ✱	1 per student	yes: collate, paper clip
TIME Tabs (sticky)	C5	63	single	1 per student	yes
Human Scale Ruler	C4	64	single	(part of TIME Tabs)	no
TIME Tabs (cutout)	C5	64	single	1 per class	yes

Notice that you **must** photocopy some pages double (front and back) and other pages single. Other pages are optional, with our recommendation listed first.

With the exception of the 9 student activity sheets (always on the left when this book is open), all other pages with recommended double-sided photocopies come printed front and back. This means you can simply remove these pages from this book and run them through a photocopier that handles multiple double pages. To this end, we provide instructions below for removing these pages.

If you're working with a photocopier limited to single-sided images, you won't have to pull the book apart. But you will have to run the front/back copies through twice, paying careful attention to which side of the paper gets printed in which orientation. If you're like us, you'll sacrifice a few pages to trial-and-error adjustments before getting it right.

> ✱ **NOTE:**
> **Pay particular attention to alignment on the Book of Scale pages:** the square grids should be very closely aligned on the two sides of your photocopies. Check this easily by holding a copy up to the light.

Removing pages to photocopy: Our pages are "perfect bound" like a notepad. If they are not too heavily glued, they may pull off one at a time from the back. But if they threaten to tear, you'll need to use a knife.

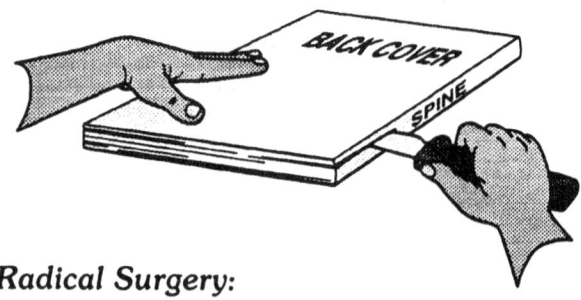

One at a time:

1. Bend the back cover all the way back and crease it (or tear it off).

2. Pull the exposed page free of the binding. Don't pull so hard that it tears. The outer few pages are often glued more securely than the rest. If the glue is too strong, you'll need to perform radical surgery (see next column).

3. Proceed to the next leaf until you remove all needed pages. (Don't try to remove pages from the middle of the book first. This only works if the binding is ready to fall apart anyway.)

Radical Surgery:

1. Place a sharp knife between pages 38 and 39 with the edge facing the binding but away from you. Close the book, and pull the knife through to cleanly divide the book into two parts.

2. Pull off the Book of Scale pages one at a time, like sheets off a notepad.

3. Repeat step 1 as needed if pages threaten to tear.

Optionally, use a sharp craft knife and a straightedge to cut pages away from the binding.

Getting Ready

❏ Decide which parts of this book you want to teach, then budget appropriate class time. See Scope and Sequence on pages 8-9 to overview a variety of options for science and math classes. There are rich possibilities for independent study and extra credit for students who are motivated to do additional, or superlative, work.

❏ Photocopy relevant student materials as per photocopy table opposite. For your convenience, copying instructions for each page are also summarized near the copyright notice at the bottom of that page.

❏ Gather these simple materials. Here is a master list of everything you'll need to teach all 9 lessons. See the teaching notes that accompany each activity for more details concerning materials that relate to specific lessons.

> ruled notebook paper
> pencils with good erasers
> calculators (scientific calculators are a nice option)
> a roll of masking tape
> scissors
> rolls of clear tape (or pre-hang strips from students' desks)
> straight pins
> several soda cans
> meter sticks
> a stapler
> metric desk rulers
> paper clips
> hand lenses (optional)
> magnetic compasses (optional)
> drawing compasses (optional)
> waxed paper (alternative to drawing compass)
> a roll of string (nonelastic kite string is ideal)

❏ Organize a way to track assignments. It may be a good idea to keep student work on file in class. If you lack file space, substitute an empty copy paper box and brick. File folders and notebooks both make suitable assignment organizers. Students will feel a sense of accomplishment as their completed papers accumulate into an impressive portfolio. Since all assignments stay together, reference and review are easy. Ask students to tape a sheet of notebook paper inside the front covers of their folders or notebooks. Track individual progress by listing and initialing lesson numbers as they are completed.

❏ Communicate your grading expectations. We recommend that you grade on individual effort, attitude and overall achievement:

✔ Effort: How many pages of work has the student produced? Of what quality?

✔ Attitude: Has the student worked to capacity, or wasted time? What evidence of personal initiative and responsibility?

✔ Achievement: Assign tasks that assess how well students have mastered key concepts. You might ask them to compare distances on their Long Tapes and Log Scales, or to complete questions in their Books of Scale as a take-home test.

Scope and Sequence

If you teach astronomy, earth/space, or general science, this book offers a high-interest, hands-on curriculum that lasts several weeks, giving your students a solid understanding of their place in the space/time universe.

If you can't spend that much time, we've divided it into independent strands A, B and C with stand-alone lessons that take way less time, down to just half of a class period. Sampling just a bit of this book now will motivate you to budget enough class time to teach more of it next year. Consider the following options to fit your available time:

math (General, Algebra)	*science* (General, Earth/Space, Physics)
1 hr: A1	0.5 hr: intro to B1
1 hr: A2	0.5 hr: intro to C5
2 hr: B1	1 – 5 hr: any math combination
1 hr: B2	4.5 hr: C1, C2, C3
2 hr: A1, A2	5.5 hr: C1, C2, C3, C4
3 hr: B1, B2	7.5 hr: C1, C2, C3, C5
5 hr: A1, A2, B1, B2	8.5 hr: C1, C2, C3, C4, C5
	11.5 hr: B1, B2, C1, C2, C3, C4, C5
	13.5 hr: A1, A2, B1, B2, C1, C2, C3, C4, C5

Section A: METRIC LONG TAPE

A1: 1 hour

SCOPE: Hike beside a LongTape through 10 orders of magnitude from 1/10 millimeter to 1,000 kilometers. Make comparisons, write metric equations, and discover qualitative and quantitative relationships along the way. Review the metric system and base-ten logic at multiple levels of understanding.

SEQUENCE: A1 stands alone with no prerequisites.

AUDIENCE: Grades 6-12 science and math.

A2: 1 hour

SCOPE: Practice the important skill of solving problems by tracking units. Wrap your mind around the spectacular difference between a thousand, million, and billion dollars in terms of stacks of $100 bills.

SEQUENCE: A2 stands alone if your students are well grounded in metrics. Otherwise do A1 first.

AUDIENCE: Grades 6-12 science and math. Use as an introduction to unit analysis (dimensional analysis) in physics and chemistry classes.

Section B: LOG SCALES

B1: 2 hours:

SCOPE: Order 36 DISTANCE Tabs from shortest to longest. Plot these on Log Scales that span 42 orders of magnitude -- from the radius of a proton to the radius of the visible universe.

SEQUENCE: B1 stands alone with no prerequisites. An answer key (pages 37-38) allows you to skip over this activity if time is short. But please consider using B1's excellent introduction to relative size (see page 15), which will spark lively classroom debate and generate many teachable moments.

AUDIENCE: Grades 6-12 general science, Earth/space science

B2: 1 hour:

SCOPE: Compare size differences by rough orders of magnitude: A pinhead is about 12 OM's larger than a proton, and roughly 12 OM's smaller than the Sun. That's a trillion times in each direction! Then compare more precisely by dividing numbers or subtracting exponents. Both methods lead to the same result.

SEQUENCE: B1 is a logical prerequisite to B2. But B2 can also stand alone if you provide an answer key (pgs 37-38). This key is also a "second chance" for those students who "messed up" B1.

AUDIENCE: Grades 6-12 general science, Earth/space science. Use B2 as a supplementary activity to the study of exponents and logarithms in algebra classes.

Section C: BOOK OF SCALE

C1: 1.5 hours:
SCOPE: Make a Book of Scale with scaled centimeter grids that "grow" or "shrink" by powers of ten with each turn of the page. Begin by drawing familiar objects at human scale – your classroom, your upper torso, your hand, a paper clip, a pin point.
SEQUENCE: C1 is a necessary prerequisite to all C-strand activities. If you decide to skip these, the introduction to C5 may still be taught in isolation, without making a Book of Scale. Students organize themselves by relative time, as they did with relative distance in the introduction to B1 (see page 27).
AUDIENCE: Grades 6-12 general science, Earth/space science. The Book of Scale students construct here is also useful in biology and astronomy classes for drawing and comparing objects of study. At the end of the course, use it as an assessment portfolio.

C2: 1.5 hours:
SCOPE: Draw to scale and label 36 DISTANCE Tabs in your Book of Scale, beginning with a proton and ending with the visible universe. Draw circles with a pair of compasses, or with a simple circle tool improvised from a strip of waxed paper and a pin.
SEQUENCE: C2 is a prerequisite to C3, C4 and C5. It fills out the Book of Scale.
AUDIENCE: Grades 6-12 general science, Earth/space science.

C3: 1.5 hours:
SCOPE: Rescale objects in your Book of Scale ten times larger or smaller on adjacent grids. Detail the solar system at 7 orders of magnitude, beginning with Mercury's orbit, through Pluto and the Kuiper Belt, out to the edge of the Oort Cloud.
SEQUENCE: C3 is a prerequisite to C4 and C5. It refines the Book of Scale, logically connecting adjacent pages.
AUDIENCE: Grades 6-12 general science, Earth/space science.

C4: 1 hour:
SCOPE: Use the Human Scale Ruler to frame abstract distances in human terms. Assuming level, unobstructed walking, discover that you can hike the distance from sea level to the top of Mt. Everest in under 2 hours; to the orbiting GLAST telescope in 11 days; to the moon in 21 years; to the Sun in 150 lifetimes.
SEQUENCE: C4 provides an optional overview of distance relationships. If time is short, you might end here and not take up the comparative study of time in C5. Alternatively, you can save an hour by skipping C4 and proceeding directly to C5.
AUDIENCE: Grades 6-12 general science, Earth/space science.

C5: 3 hours:
SCOPE: Order 36 TIME Tabs from shortest to longest. Tape them to "light clocks" at the bottom of each grid in your Book of Scale, calibrated in time spans ranging from much shorter than a nanosecond, to way longer than a light year. Interact with a wide range of creative problems that help students better understand the space-time continuum.
SEQUENCE: C5 provides a broad summary of scale and orders of magnitude across space and time. It draws together important ideas in the Book of Scale, and provides a platform for further open-ended independent study.
AUDIENCE: Grades 6-12 general science, Earth/space science.

Reproducible
STUDENT ACTIVITY PAGES
with
TEACHING NOTES

ORDERS OF MAGNITUDE

Activity A1

1. Get a *LongTape* page. Imagine taking a journey that begins at zero millimeters on this tape, and ends 620 miles (1 megameter) above Earth, twice as high as NASA's* orbiting gamma-ray telescope, called GLAST**. Write a story about your adventures. It might begin like this, or in any way you like:

*NASA: The National Aeronautics and Space Administration
**GLAST: Gamma-ray Large Area Space Telescope

My Metric Journey

Well, by taking only one step, I passed these 4 orders of magnitude, each one ten times farther from my starting point:
0.1 mm = 10^{-4} m, about as thin as paper money,
1 mm = 10^{-3} m, about as thin as a dime,
1 cm = 10^{-2} m, about as thick as a big fat pencil,
10 cm = 10^{-1} m, about wide as my hand.
Even an ant would have reached those markers really fast! I now find myself leaving behind the 1 meter marker, which reads:
1 m = 10^{0} m ~ a giant step.
With 5 markers down and only 6 more to go, it feels like this trip won't last long! My progress has been remarkable so far. Already, I'm 10,000 times farther along than when I passed the first marker. And the 6th marker is already in sight, it's just over the hill....

2. Compare these units mm, cm, m, km and Mm as follows:

a. Copy these METRIC COMPARISONS on lined paper, then extend your list. Vary your phrasing as underlined:

1 m is 100 <u>times</u> longer than 1 cm.
1 km is 3 <u>powers of ten</u> shorter than 1 Mm.
1 m is 3 <u>orders of magnitude</u> longer than 1 mm.

b. Copy these EQUATION PAIRS and extend your list. Write numbers in different ways, using decimals, powers of ten or words as shown.

1 km = 1,000 m (and) 1 m = 0.001 km
10^0 cm = 10^{-2} m (and) 10^0 m = 10^2 cm
one mm = one millionth km (and) one km = one million mm

c. List INTERESTING FACTOIDS. Start with these and extend your list.

10,000 fat pencils placed side by side reach goal to goal on a soccer field.
A car traveling 100 km/hr takes .00001 hours to travel 1 meter.
A 1 mm stack of hundred dollar bills is worth $1,000.

OVERVIEW / OBJECTIVES

To review metric prefixes, visualize metric distances, and explore decimal relationships on a "meter-tape" graphic that spans 10 orders of magnitude.

TIME: 1 hour.

INTRODUCTION

Photocopy and distribute the LongTape supplement.

★ Count up *and* down the LongTape in different ways, depending on the level of your students. Always begin at the **meter** marker:

- *Say:* **1 m**, 10 m, 100 m, **1 km**, 10 km, 100 km, **1 Mm**.
- *Say:* **1 m**, 10 cm, 1 cm, **1 mm**, 0.1 mm.
- *Say:* giant step, length of a classroom, length of soccer field, ...
- *Say:* giant step, hand's width, wide as a fat pencil, ...
- *Say:* **one**, ten, hundred, **thousand**, ten thousand, hundred thousand, **million**, ... (keep going) ..., ten million, hundred million, **billion**, ten billion, hundred billion, **trillion**.
- *Say:* **one**, tenth, hundredth, **thousandth**, ten thousandth, ... (keep going) ..., hundred thousandth, **millionth**, ten millionth, hundred millionth, **billionth**, ten billionth, hundred billionth, **trillionth**.
- *Write:* **1**, 10, 100, **1000**, ... (Continue to a trillion.)
- *Write:* **1**, 0.1, 0.01, **0.001**, 0.0001, ... (to a trillionth)
- *Say:* **1**, 10, 10×10, **10×10×10**, 10×10×10×10 ...
- *Say:* **1**, 1/10, 1/10/10, **1/10/10/10**, ...
- *Write:* 10^0, 10^1, 10^2, 10^3, ... (to a trillion)
- *Write:* 10^0, 10^{-1}, 10^{-2}, 10^{-3}, ... (to a trillionth)
- *Say:* **0 orders of magnitude** larger, 1 order of magnitude larger, **2 OM's** larger, **3 OM's** larger, ...
- *Say:* **0 OM's** smaller, 1 OM smaller, 2 OM's smaller, **3 OM's** smaller, ...
- *Write:* **1 m**, 10 m, 100 m, **1000 m**, ...
- *Write:* **1 m**, 1/10 m, 1/100 m, **1/1000 m**, ...

★ Convert decimals and powers of ten:

- You say the number and students say the power of ten.
  - one = "ten to the zero power"
  - billion = "ten to the ninth power"
  - ten thousandth = "ten to the minus four power"
- You say the power of ten and students say the number.
  - ten to the first power = "ten"
  - ten to the sixth power = "million"
  - ten to the minus two = "hundredth"

★ Place paper clips on any two OM markers. Write equalities in scientific notation and decimal forms as follows.

EXAMPLE 1: Markers placed on 1 kilometer and 1 meter are separated by 3 orders of magnitude. Thus...

1 km = 10^3 (1 m)		1 m = 10^{-3} (1 km)
1 km = 1 × 10^3 m	*and*	1 m = 1×10^{-3} km
1 km = 1,000 m		1 m = 0.001 km

In all of these equations, notice that...

One big = many small
AND
One small = a fraction of big

EXAMPLE 2: Markers on 10 kilometers and 1 centimeter are separated by 6 orders of magnitude.

10 km = 10^6 (1 cm)		1 cm = 10^{-6} (10 km)
10 km = 1 × 10^6 cm	*and*	1 cm = 10 × 10^{-6} km
1 km = 1 × 10^5 cm		1 cm = 1 × 10^{-5} km
1 km = 100,000 cm		1 cm = 0.00001 km

NOTES / ANSWERS

Steps 1 and 2 serve as stand-alone, open-ended reviews of the metric system that all students can use at multiple levels of understanding. Consider assigning these for a specified time period, say one class period plus homework. Students might explore some questions in depth, or choose to answer all questions in less depth. Ultimately, each learner is responsible for his or her own learning. Allow students to exercise their muscles of choice and responsibility whenever possible.

1. Here the "language arts majors" in your class have a chance to shine. Consider pairing them with students who wield numbers with more confidence. Assign these teams to create stories for a travel magazine.

Encourage students to use both imagination and math skills as they develop this story. If it takes 12 minutes to walk 1 kilometer, for example, when multiplying by powers of ten, it takes (10)(10)(10)(12 minutes) = 200 hours to walk a megameter (if that distance were on level ground).

2a,b. Here is a summary of metric distance relationships:

1 mm: 10^{-1} cm, 10^{-3} m, 10^{-6} km, 10^{-9} Mm
1 cm: 10^1 mm, 10^{-2} m, 10^{-5} km, 10^{-8} Mm
1 m: 10^3 mm, 10^2 cm, 10^{-3} km, 10^{-6} Mm
1 km: 10^6 mm, 10^5 cm, 10^3 m, 10^{-3} Mm
1 Mm: 10^9 mm, 10^8 cm, 10^6 m, 10^3 km

2c. A few more interesting "factoids." (Students may list others.):

- The difference between a **million** dollars and a **billion** dollars is like the difference between a **meter** stack of $100 bills and a **kilometer** stack!
- A ten kilometer stack of dimes is worth $1 million.
- A jogger runs 10,000 meters/hour or 2.8 meters/sec.
- A jogger runs 1 meter in 0.0001 hours.

MATERIALS

☐ Photocopy the LongTape supplement on page 32.
☐ Notebook paper, pencil and eraser. These common student materials are assumed in all later lessons.
☐ Calculators (optional).

NOTES: **Activity A1**

UNIT ANALYSIS

Activity A2

1. A $100 bill measures 0.1 mm thick on your *LongTape*. So, ten of them stacked together ($1,000) is 1 mm thick, and $10,000 reach 1 cm...

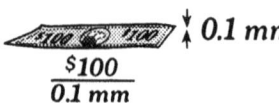
$\frac{\$100}{0.1\ mm}$

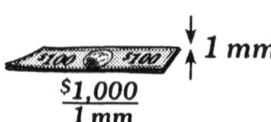
$\frac{\$1,000}{1\ mm}$

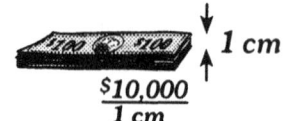
$\frac{\$10,000}{1\ cm}$

a. Label the **cash value** of each distance marker on your *LongTape* for a stack of $100 bills that high. Write in the margins.

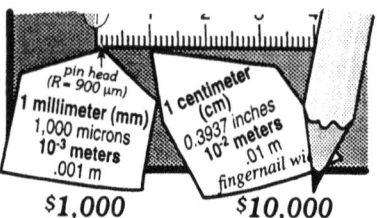

b. Complete this series of conversion factors.

$$1 = \frac{\$1,000}{mm} = \frac{\$10,000}{cm} = \frac{?}{m} = \frac{?}{km} = \frac{?}{Mm} = 1$$

---

2. How high will $100 bills stack for the following annual incomes? Use your dollar-to-distance table, plus these additional **unit conversion factors**, to find out.

Work Time Table *(Standard working hours)*

8 hours/day	5 days/week	48 weeks/year

Work Time Table *(Rat Race hours)*

16 hours/day	6 days/week	52 weeks/year

*You can use **UNIT ANALYSIS** to answer the questions below.*

EXAMPLE

a. How high a stack for the annual income of an **undocumented worker** holding down 2 jobs and working Saturdays. (Figure $5/hr):

$$\frac{\$5}{1\ hr} \times \frac{16\ hr}{1\ day} \times \frac{6\ day}{1\ week} \times \frac{52\ week}{1\ year} \times \frac{1\ cm}{\$10,000} \approx \frac{2.5\ cm}{1\ year}$$

Now try these:

b. Annual income if you quit school now and work *standard* hours for **minimum wage**.
c. Annual income for working *standard* hours at a high-paying union factory job. ($30/hr)
d. Annual income you can **realistically expect** to earn after graduating from high school or college. Plug in your most likely hours and wages.
e. Annual income from a **dream job**. Plug in the hours and wages that satisfy your sense of "quality of life" and "enough."
f. Annual income of an **average CEO** trying to win the rat race. (Figure $600/hr).

---

3. Find the height of each stack of cash in $100 bills.

EXAMPLE

a. GLAST total budget... ($350 million)

$$\text{GLAST budget} = \frac{\$350\ million}{1} \times \frac{1\ meter}{\$1\ million}$$
$$= 350\ meters = 0.35\ km$$

Now try these:

b. U.S. deficit spending in 2004 ... ($521 billion)
c. U.S. national debt at end of 2004 ... ($7.2 trillion)
d. U.S. gross national product in the year 2004 ... ($10.5 trillion)

---

4. Start with the left side of each equation. Find conversion factors on your *LongTape* that multiply together to give you what's on the right side.

EXAMPLE

Map it first: yard to 100 m to feet.

Then set up **unit fractions** and plug in the numbers.

$$\frac{1\ yard}{1} \times \frac{100\ m}{109\ yard} \times \frac{32.8\ feet}{10\ m} = 3.01\ feet$$

(This answer is close to 3, but not exact because conversion factors are rounded off.)

Now try these:

b. 1 yard = 36 inches
  Map: yard to m to inches.
c. 1 mile = 1,760 yards
d. 1 foot = 0.333 yards
e. 1 mile = 5,280 feet
f. 1 foot = 12 inches
g. make your own comparison

OVERVIEW / OBJECTIVES

To express equations as unit conversion factors. To use unit analysis in problem solving.

TIME: 1 hour.

INTRODUCTION

★ Students are familiar with **n**ickels, **d**imes and **q**uarters. Use these relationships to teach **unit analysis**. First ask volunteers to express the mathematical equalities between these three coins:

$$2n = 1d, \quad 5n = 1q, \quad 2.5d = 1q$$

Anything divided by itself equals "1." So the left side of each equation divided by its right side equals one:

$$1 = 1d/2n = 1q/5n = 1q/2.5d$$

Equally, the right side of each equation divided by its left side creates an additional trio of inverted fractions:

$$1 = 2n/1d = 5n/1q = 2.5d/1q$$

All of these different ways of writing "one" are called **unit conversion factors**. They are useful for solving all kinds of math and science problems. Try this example:

$$4n = ?d$$

Strategy: Write what's given as a fraction (with a denominator of 1). Then multiply by a conversion factor that cancels the unit you are given, to leave you with the unit you want. In this case, choose any conversion factor with n in the denominator. These conversion factors do the job:

$$\frac{4n}{1} \times \frac{d}{2n} = 2d \quad \text{(and)} \quad \frac{4n}{1} \times \frac{q}{5n} \times \frac{2.5d}{q} = 2d$$

Apply unit analysis to these (or other) simple coin problems until your students become familiar with the process. This is a valuable skill well worth spending time on.

$$10n = ?q, \quad 4q = ?d, \quad 3d = ?n, \quad 3q = ?n, \quad 5d = ?q$$

★ Why study unit conversions? What's the point? Problems in physics, chemistry, and sometimes life, typically give you numbers in one unit and ask for solutions in another unit. Solving such problems means choosing unit conversion factors that, when multiplied, cancel to what you want. Learn this skill and become a better problem solver!

MATERIALS

☐ LongTape supplement.
☐ Advise students of the current minimum wage.
☐ A calculator.

NOTES / ANSWERS

As before, consider assigning students to work on this activity over a specified period of time. Devoting some effort and attention to each of the 4 numbered problem areas may be more important than answering all multiple lettered parts.

1a. Students should indicate the cash value of each distance marker on their LongTape in terms the value of a stack of hundred dollar bills that high:

0.1 mm = $1 hundred	1 m = $1 million	1 km = $1 billion
1 mm = $1 thousand	10 m = $10 million	10 km = $10 billion
1 cm = $10 thousand	100 m = $100 million	100 km = $100 billion
10 cm = $100 thousand		1 Mm = $1 trillion

1b. $1 = \dfrac{\$1{,}000}{1\text{ mm}} = \dfrac{\$10{,}000}{1\text{ cm}} = \dfrac{\$1\text{ mil}}{1\text{ m}} = \dfrac{\$1\text{ bil}}{1\text{ km}} = \dfrac{\$1\text{ tril}}{1\text{ Mm}}$

2b. Answers will vary depending on state's minimum wage. Example for $7.50 minimum wage:

$$\frac{\$7.50}{1\text{ hr}} \times \frac{8\text{ hr}}{1\text{ day}} \times \frac{5\text{ day}}{1\text{ week}} \times \frac{48\text{ week}}{1\text{ year}} \times \frac{1\text{ cm}}{\$10{,}000} \approx \frac{1.44\text{ cm}}{1\text{ year}}$$

2c. $\dfrac{\$30}{1\text{ hr}} \times \dfrac{8\text{ hr}}{1\text{ day}} \times \dfrac{5\text{ day}}{1\text{ week}} \times \dfrac{48\text{ week}}{1\text{ year}} \times \dfrac{1\text{ cm}}{\$10{,}000} \approx \dfrac{5.76\text{ cm}}{1\text{ year}}$

2d, e. Answers will vary.

2f. Answers will vary with interpretations of amount of work required to compete in "the rat race." An example:

$$\frac{\$600}{1\text{ hr}} \times \frac{10\text{ hr}}{1\text{ day}} \times \frac{6\text{ day}}{1\text{ week}} \times \frac{48\text{ week}}{1\text{ year}} \times \frac{1\text{ cm}}{\$10{,}000} \approx \frac{172.8\text{ cm}}{1\text{ year}}$$

or 1.728 m/yr

3b. $\dfrac{\$521\text{ billion}}{1} \times \dfrac{1\text{ km}}{\$1\text{ billion}} = 521\text{ km}$

3c. $\dfrac{\$7.2\text{ trillion}}{1} \times \dfrac{1\text{ Mm}}{\$1\text{ trillion}} = 7.2\text{ Mm} = 7{,}200\text{ km}$

3d. $\dfrac{\$10.5\text{ trillion}}{1} \times \dfrac{1\text{ Mm}}{\$1\text{ trillion}} = 10.5\text{ Mm} = 10{,}500\text{ km}$

4b. $\dfrac{1\text{ yd}}{1} \times \dfrac{100\text{ m}}{109\text{ yd}} \times \dfrac{39.37\text{ in}}{1\text{ m}} = 36.1\text{ in}$

4c. $\dfrac{1\text{ mi}}{1} \times \dfrac{1\text{ km}}{0.62\text{ mi}} \times \dfrac{1000\text{ m}}{1\text{ km}} \times \dfrac{109\text{ yd}}{100\text{ m}} = 1758\text{ yd}$

4d. $\dfrac{1\text{ ft}}{1} \times \dfrac{10\text{ m}}{32.8\text{ ft}} \times \dfrac{109\text{ yd}}{100\text{ m}} = 0.332\text{ yd}$

4e. $\dfrac{1\text{ mi}}{1} \times \dfrac{1\text{ km}}{0.62\text{ mi}} \times \dfrac{1000\text{ m}}{1\text{ km}} \times \dfrac{32.8\text{ ft}}{10\text{ m}} = 5290\text{ ft}$

4f. $\dfrac{1\text{ ft}}{1} \times \dfrac{10\text{ m}}{32.8\text{ ft}} \times \dfrac{39.37\text{ in}}{1\text{ m}} = 12.00\text{ in}$

4g. Other possibilities:
- Express feet, years or miles in terms of inches.
- Convert walking or driving speeds defined by the LongTape into feet/second.
- Express a small unit in terms of a larger unit, such as an inch as a fraction of a mile.

NOTES: **Activity A2**

ORDERING DISTANCE

Activity B1

1. Get the *Distance Tabs*. Compare this actual ordering to what you decided as a class. For each category, write about what surprised you and what you learned.

2. Get the *Log Scales* (2 pages). Tape your *Distance Tabs* to these scales as follows:

  a. Cut out the "Human Scale" strip of tabs like a "picket fence."

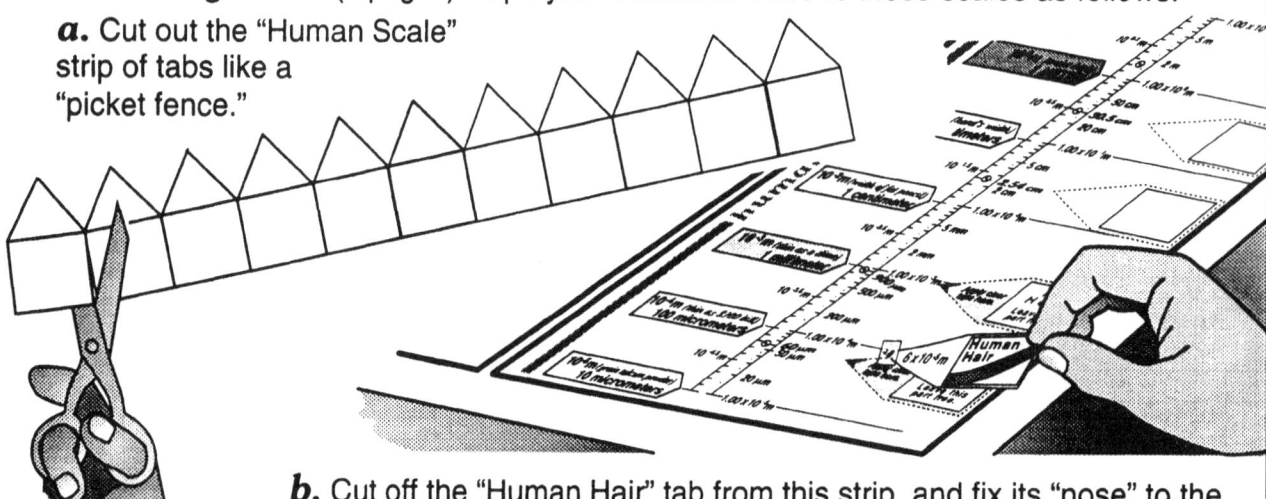

  b. Cut off the "Human Hair" tab from this strip, and fix its "nose" to the shaded "point" on the *Log Scales* with a bit of clear tape.

  c. Place the "Pin Head" tab next. Notice that its radius (expressed in both scientific notation and as a base-ten exponent) is **plotted** and **labeled** in local units on the *Log Scale*.

  d. Continue in like manner (**cut** tab, **plot** point, **label** unit, **tape** tab), until you accurately place all tabs along both *Log Scale* pages. (Note: Tabs may overlap. Tape only the noses down so you can lift tabs to see underneath.)

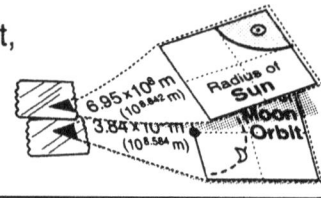

3. Use the front section of your *Long Tape* like a ruler to measure, plot and label these additional distances:

  ✔ **Diameter of a pin hole.** (Poke actual pinholes through the paper. Examine with a magnifier to estimate width.)

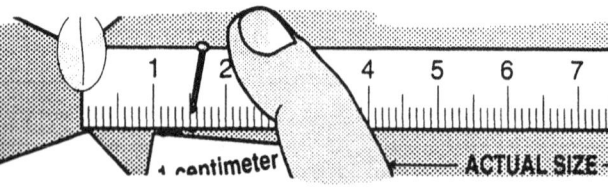

  ✔ Diameter of your Pencil.
  ✔ Diameter of a Soda Can.

4. Measure these distances with a meter stick. Plot and label them on your *Log Scale*.

  ✔ Length of your Hand Span.
  ✔ Length of your Arm.
  ✔ Length of your Room.

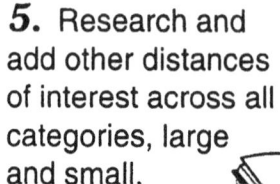

5. Research and add other distances of interest across all categories, large and small.

OVERVIEW / OBJECTIVES

To qualitatively compare and sort distances, from submicroscopic to astronomical, and plot these on a log scale. To work cooperatively and individually toward a more accurate understanding of how structures in the universe fit together.

TIME: 2 hours.

INTRODUCTION

★ This distance-ordering exercise is a necessary, and fascinating, introduction to activity 3. You'll need 1 copy (only) of the Distance "Sticky Tabs" found on page 33. Cut and tape each group (there are 4), as directed on this page. Follow these 4 steps for each category, beginning with the "*human scale*" group:

1. Distribute 1 tab per student. (Up to 36 students per class may eventually participate.) In classes of 12 or less, students may participate multiple times and sort more than one tab at a time.

2. Ask students to **read** their tab(s) aloud and **discuss** the described distance. (See concepts key below, divided by category.)

3. Have students **sort** themselves in a line from the one with the shortest tab to the longest. If students have multiple tabs, they can stick them along a table or wall.

4. Review the ordering. Does everyone agree that this is longer than that? Play dumb, questioning correct as well as incorrect orderings. Allow at least a few mistakes to stand so students will have something to write about in question 1. Have students pass single file, sticking their tabs along a table or wall in the same line order for later reference.

KEY CONCEPTS: human scale

- Altitude of **GLAST Orbit**: GLAST is an Earth-orbiting space telescope falling continuously around the Earth's curvature about 550 km above sea level.
- Distance **Sound** Travels: Imagine how far you must stand from a person to *see* them clap their hands together, then *hear* that sound 1 second later.

subatomic/microscopic

- Recall that an **atom** has a small, dense **nucleus** composed of neutrons and **protons**.

solar system and stars

- The **radius** of a circle or sphere is half its diameter.
- Most tabs involve radii, but some are **distances** from you **to** something else.
- Some tabs describe **physical objects**; others **orbits**.
- Radius of **Oort Cloud**: Its outer boundary defines the end of the Sun's gravitational sphere of influence; the origin of long-period comets.

astronomical

- Gravity holds our universe together in a complex dance of objects orbiting objects orbiting objects...
- **Earth** orbits our home star (the **Sun**).
- The **Sun** orbits the center of our **Milky Way Galaxy** along with billions of other stars (including **Sirius**, **HD70642**, and a spinning, collapsed stellar core called the **Crab Pulsar**.)
- Our **Milky Way** plus smaller satellite galaxies (like the **Large Magellanic Cloud**) orbit around numerous other more distant galaxies (like **Andromeda**), to form a **Local Group**.
- Our **Local Group** rotates near the outer edge of many other groups of galaxies with the **Virgo Group** near its center. This huge, gravitationally bound "group of galaxy groups" is called the **Virgo Supercluster**.
- The **Virgo Supercluster** floats like a dust mote among the billions of other superclusters that comprise the **Observable Universe**.
- A good question to ask: Is this object *inside* the Milky Way, or many orders of magnitude more distant?
- **GRB 990123**: This is not an object, but rather **an event** — a Gamma Ray Burst — that occurred on January 23, 1999. Something flashed briefly with the energy of a billion billion Suns, then slowly faded away. Was this the birth of a black hole?
- **AGN 3C 273**: This Active Galactic Nucleus has an energy of a trillion suns (10^{12}), which is a million times less than a billion billion suns (10^{18})! But, this energy is *sustained* at the center of a galaxy over billions and billions of years!

NOTES / ANSWERS

The introduction will spark class discussion and debate that could last an entire class period. Plotting Distance Tabs along the Log Scales requires an additional hour. Question 5 might be assigned as homework.

2. Students should cut just one tab off the "picket fence" at a time, then plot and tape it to the Log Scale before cutting off the next one, thus preserving their ordered arrangement.

Suggest that students tape only the "nose" of each tab to the Log Scale (beside their plotted points) using short snippets of clear tape. (You can distribute longer strips in advance, hanging them on the edge of each desk.) Tape the tabs far enough to the right that they don't cover numbers on the scale. Don't use glue – layered tabs may bond into unreadable lumps.

Notice that each tab distance is defined in **scientific notation** (which students can plot relative to calibrations on the **left**) and as a **base-ten exponent** (which students can plot relative to calibrations on the **right**). Either way, the point ends up in the exact same place. Insist that students include "local units" (nanometers, kilometers, light years, etc) with each plotted point. Later on, these units will help students decide how large they should make their scale drawings.

Find a complete answer key on pages 37-38.

MATERIALS

- ☐ DISTANCE Tabs (sticky), pg 33, 1 single-side copy.
- ☐ DISTANCE Tabs (cutout), pg 34, 1 single-side copy per student.
- ☐ Log Scales, pgs 35-36, 1 double-sided copy per student.
- ☐ Scissors.
- ☐ Clear tape.
- ☐ Straight pin, pencil, and soda can.
- ☐ The Long Tape supplement.
- ☐ Hand lens (optional).
- ☐ A meter stick.

NOTES: **Activity B1**

USING A LOG SCALE

Activity B2

1. Compare arrows *a* through *e* to 1 METER! Vary the way you say things. Don't get stuck in a rut.

a. EXAMPLE: Different ways to compare **a femtometer**. (Choose one.)

A *femtometer* is 15 orders of magnitude smaller than a *meter*...
A *femtometer* is 10^{15} X smaller than a *meter*...
A *femtometer* is 1,000,000,000,000,000 X smaller...
A *femtometer* is a million billion X smaller...
A *femtometer* is a thousand trillion X smaller...
10^{15} femtometer = 1 meter...
1 femtometer = 10^{-15} meter...

e. EXAMPLE: Different ways to compare the **thickness of a human hair**. (Choose one.)

A *human hair* is more than 4 OM's thinner than a *meter*...
A *human hair* is more than 10^4 X thinner...
A *human hair* is more than 10,000 X thinner...
A *human hair* is more than ten thousand X thinner...
A *human hair* is $10^{4.222}$ X thinner...
A *human hair* is 4.222 orders of magnitude thinner...
A *human hair* is about 17,000 X thinner...

2. Distances in each triplet are separated by the same **rough order of magnitude**! Find each OM, then write a wonderful comparison.
EXAMPLE: *a.* 1 OM: Earth is to Jupiter as Jupiter is to the Sun. This means about 100 Earths span 1 Sun!

    a. Radius of **Earth** → Radius of **Jupiter** → Radius of **Sun**
    b. Length of **Bacterium** → Height of **Human** → Radius of **Moon**
    c. Radius of **Proton** → Radius of **Pin Head** → Radius of **Sun**
    d. Thickness of 2 **hairs** → 1 **Astronomical Unit** → Radius of **Universe**

3. Compare these distances more precisely by dividing numbers or subtracting exponents. Try these, then some of your own.

    a. avg virus → avg bacterium   *c.* human → Mt. Everest
    b. proton → hydrogen atom   *d.* to Sirius → to Andromeda

EXAMPLE: *a.* Divide shorter length (the virus) into the longer (the bacterium)...

using numbers:
$2 \times 10^{-6} m / 7.5 \times 10^{-8} m =$
$0.266 \times 10^2 = 27 \times$ longer.

using exponents:
$10^{-5.699} / 10^{-7.125} =$
$10^{1.426} = 27 \times$ longer.

4. At 10^0 m you stride through your world in meters. Imagine shrinking 100 X and walking in centimeters, then another 10 X smaller to take tiny millimeter steps....

    a. Describe your adventures.
    b. Imagine cutting a meter stick with an infinitely sharp knife. Can you end up with *no* meter stick by moving down the *Log Scale*?

5a. Complete these patterns of equations to two significant figures on a scientific calculator:

    b. Relate these figures to the Log Scale interval between 1m and 10m.
    c. Investigate other intervals. How are Log Scales calibrated?

$10^{1.0} = 10$	$\log 10 = 1$
$10^{0.9} = 7.9$	$\log 9 = .95$
⇓	⇓
$10^{0.1} =$	$\log 2 =$
$10^{0.0} =$	$\log 1 =$

6. Compare your *Long Tape* and *Log Scale*. List advantages and disadvantages of these ways to calibrate distance.

OVERVIEW / OBJECTIVES

To compare distances by orders of magnitude. To make these comparisons by dividing numbers and by finding the difference between base-10 exponents.

TIME: 1 hour.

NOTES / ANSWERS

Again, you might give students some choice in problem selection. Numbers 4 and 6 exercise communication skills, while 3 and 5 exercise advanced math skills.

1a-d. These problems ask students to compare whole orders of magnitude. (The diameter of a grain of talcum powder is listed as 10^{-5} m on the log scale.)

1e-g. These problems ask students to compare fractional orders of magnitude at various levels of math expertise.

2a. 1 OM: Earth is to Jupiter as Jupiter is to the Sun. This means about 100 Earths span 1 Sun! *(Since volume varies with the cube of distance, roughly a thousand Earths fill one Jupiter, and a million Earths fill one Sun.)*

2b. 6 OM's: You exploring the moon is like a bacterium exploring your skin! You're a million times smaller than the moon, and a million times bigger than the bacterium.

2c. 12 OM's: A pinhead is roughly a trillion times smaller than the sun, and a trillion times larger than a proton.

2d. 15 OM's: If you were to shrink our visible universe by a million billion times, it would roughly fit Earth's orbit around the Sun. At this scale, the Earth-Sun distance (1 AU) would shrink to about 2 hairwidths.

3a. *See example given with problem.*

3b. hydrogen atom / proton =
5.29×10^{-11} m / 8.7×10^{-16} m = $0.608 \times 10^5 \approx 60{,}800 \times$ longer.
$10^{-10.277} / 10^{-15.060} = 10^{1.426} \approx 60{,}800 \times$ longer.

3c. Mt. Everest / human =
8.85×10^3 m / 1.7×10^0 m = $5.2 \times 10^3 \approx 5{,}200 \times$ longer.
$10^{3.947} / 10^{0.230} = 10^{3.717} \approx 5{,}200 \times$ longer.

3d. to Andromeda / to Sirius =
2.9×10^{22} m / 8.6×10^{16} m = $0.337 \times 10^6 \approx 337{,}000 \times$ longer.
$10^{22.462} / 10^{16.934} = 10^{5.528} \approx 337{,}000 \times$ longer.

3. Other interesting comparisons:
average human / one foot $\approx$ 5.6 times (or 5.6 feet)
length of basketball court / average human $\approx$ 15 times.
(Check this out in the gym with 15 student volunteers.)
observable universe / proton = $1.6 \times 10^{41} = 10^{41.206}$.
(So a hundred thousand trillion trillion trillion protons span the universe, and 10^{124} protons fill the sphere.)

4a. *A sample story starter:* Using a sophisticated virtual travel device, (attached by electrodes to my skull) I can shrink at will (along with my device) to smaller orders of magnitude. Setting the dial to 0, I walk in meter strides around the room (10^0 m steps). My normal step is something less than a meter, yet this pace feels quite natural. Odd, though, everything around me now seems a bit small. And surprisingly, I'm the tallest kid in class! I thought this device was supposed to make me smaller! Cautiously I turn the dial to -1...

4b. If my knife is infinitely sharp, I will always be able to cut off nine tenths of the remaining meter stick, no matter how minute that length becomes.

5a.

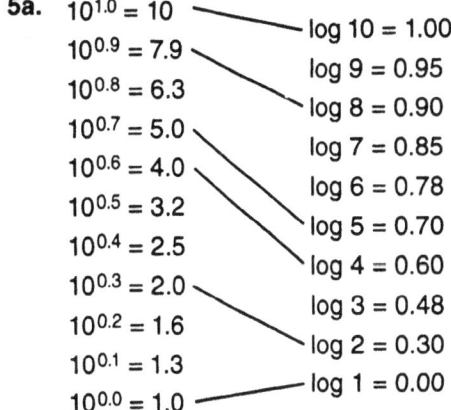

5b. Left column: The 1-10 interval is calibrated evenly up the **left** side in **exponents**: 0.0, 0.1, 0.2, ..., 1.0. When 10 is raised to these powers on a calculator, you get a set of corresponding **numbers** calibrated up the **right** side with uneven steps (expanded at the bottom and bunched at the top.)

Right column: Taking the base 10 logs of the unevenly spaced **numbers** on the **right** produces the evenly spaced **exponents** on the **left**. (Six places where exponent marks and whole number marks meet are indicated by the connecting lines above.)

5c. This correspondence between even exponents and bunching numbers holds over all intervals of the log scale. For example:
$10^{0.3} = 2$, $10^{1.3} = 20$, $10^{2.3} = 200$, $10^{3.3} = 2000$
log 2 = 0.3, log 20 = 1.3, log 200 = 2.3, log 2000 = 3.3

So log scales are calibrated (evenly) in exponents, but labeled (unevenly) in whole numbers that bunch and expand to preserve equality between both systems.

6. The LongTape is calibrated evenly for its whole length in meters (with subdivisions), starting at zero. So it is good for *measuring* distance. The LongTape spans ten orders of magnitude conceptually, but a real measuring instrument, such as the traditional meter stick, might span only five: m, 10 cm, cm, mm, and 0.1 mm (estimated between mms.)

The Log Scale has no zero. It is unbounded in both directions, good for *representing* distance across unlimited orders of magnitude. Since equal space is assigned to each power of ten, small measures don't shrink to invisibility, nor do large measures expand to physically unmanageable distances.

MATERIALS

☐ Log Scales. These were plotted and tabbed in the previous activity. Or you can photocopy the answer key on pgs 37-38, 1 double-sided copy per student who needs it.

☐ A scientific calculator to evaluate powers of 10.

NOTES: **Activity B2**

SCALE THE UNIVERSE (1)

Activity C1

1. Get 12 *Book of Scale* pages. Carefully cut each individual sheet along the middle dashed line, creating 24 half-pages.

a. Assemble these pages in order, from **1 of 24** on top, to **24 of 24** on the bottom.

b. With the *front* toward you, tap the bottom and right edges even. Staple where indicated, twice on the front, and once on the back.

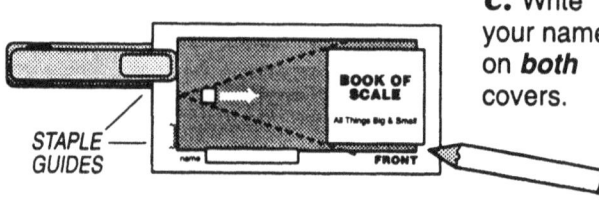

c. Write your name on **both** covers.

2. You can draw *any* object in the universe, large or small, in this *Book of Scale*! Start by plotting your **classroom** on the page with the scale of **1 cm : 10^0 m**.

a. First sketch a rough bird's eye view of your room on scratch paper. Include...
- the walls, doors and windows,
- the position of your desk and chair,
- this opened *Book of Scale* resting on your desk,
- other features of interest.

b. Use a meter stick to take all necessary measurements. Record these numbers on your sketch in meters.

c. Draw an arrow on your sketch that points North.

d. Now draw your sketch neatly to scale on the 10^0 m grid where each cm represents 1 meter. Orient your drawing (if convenient), so North points toward the top of the page.

> *Drawing Guidelines:*
> 1. Use a sharp pencil with a good eraser.
> 2. Draw straight lines with a ruler or straightedge.
> 3. Label objects in your scale drawings.
> 4. Be neat and accurate.
> 5. Stay on the grid. (Outside questions come later.)

3. 10^{-1} m: Draw **yourself**, waist up, as if you are looking out of a one-square-meter window. Begin with a rough sketch, as before. Work with a partner in taking measurements.

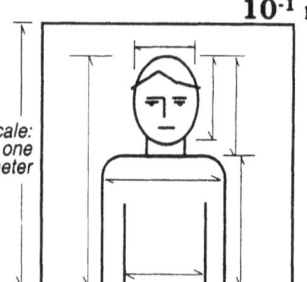

4. 10^{-2} m: Trace your **fingers**, plus as much hand as will fit, onto this *actual size* 10 cm grid. Add a **straight pin** and a **paper clip**, also actual size.

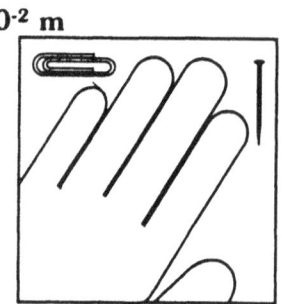

5. 10^{-3} m: Hold the 2-loop end of a **paper clip** on the smaller actual-size mm grid. Draw it to scale, square for square, in the larger window, with clip and wire in accurate proportion. Add enlarged side views of a **pin head** and **pin point** in a similar manner.

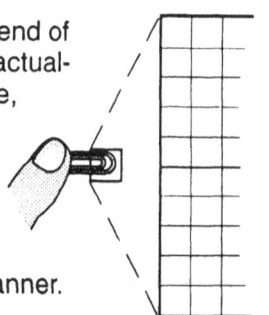

6. 10^{-4} m: Center a **pin point** on the tiny actual-size mm window and estimate what portion of the window gets covered. (Use a hand lens if available.) Now sketch this pin point 100 times larger on the larger grid.

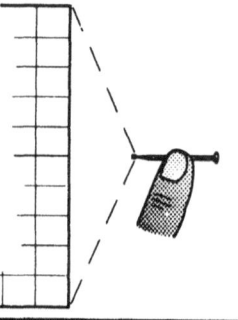

OVERVIEW / OBJECTIVES

To construct a Book of Scale with centimeter grids scaled across 44 orders of magnitude from subnuclear to astronomical. Students begin somewhere in the middle of this book by mapping their classroom and drawing themselves.

TIME: 1 hour.

INTRODUCTION

★ Make a Book of Scale to show your students what they will be assembling. *Any* object, structure or distance in the visible universe nicely fits on a page in this book!

NOTES / ANSWERS

1b. Students should firmly hold this alignment (bottom and right edges even), until stapling at the left edge makes it permanent. Supervise this step closely so the books have neat, easy-to-turn pages.

2. Orient your students to their newly-constructed Books of Scale. Question them about the properties of the room-size grid @ 10^0 meters:

What is the *actual* length of each small square? *(1 cm)*

What is the *scale* length of each small square? *(1 m)*

What scale is indicated at the top of this grid? *(1 cm represents 1 meter, or 100 cm)*

What 3 scale measurements are defined along the left side of this grid?

1 mm represents *10 cm*
1 cm represents *1 m*
10 cm represents *10 m*

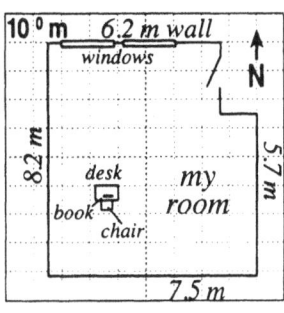

A different kind of scale called a "Light Clock" runs across the bottom of this grid. How is it calibrated? *(In units of time: nanoseconds on this page.)*

At this scale, how fast would a photon of light travel across the bottom of this grid from left to right? *(10 meters in 33.3 nanoseconds, or 0.3 m/ns.)*

Should you answer the question to the right of the grid? *(Not now. Answer it later after you have completed your Book of Scale.)*

How are neighboring pages scaled? *(The page to the left is a power of ten smaller, to the right it is a power of ten larger.)*

2b. Each square meter of floor space corresponds to a square centimeter of scaled grid space. Room measurements rounded to the nearest cm will allow scale drawings precise to the nearest 0.1 mm.

2c. If students can't locate north, observe that in the northern hemisphere, the Sun culminates due south at 12 noon, local standard time. Or provide a magnetic compass.

2d. The 10 x 10 cm grid accommodates rooms no longer (or wider) than 10 x 10 meters. If your room is larger than this, ask students to draw the quadrant of the floor plan that includes their table and chair.

Students tend to draw small objects too big. The open Book of Scale (14 cm x 40 cm), for example, should appear no larger than a fingernail clipping (just 1.4 mm x 4 mm).

Obviously, a sharp pencil, a handy eraser, and a straight-edge are important for neat and accurate work. Discuss these drawing guidelines now, as students begin this first scale drawing of many. Your expectations are most easily established in the beginning, before bad habits form.

Some young people have more artistic than scientific ability. Encourage them to draw objects with dimensional shading and exciting (but accurate) detail – little showpieces that push the limits of visual excellence. For other objects in their Books of Scale, students may want to look up photographs in the library or on the internet, and perhaps add color to their drawings. Reward them with extra credit for superior work. They will have learned a great deal in this process.

3. Some may want to spend a lot of time on this ego-sensitive self portrait, trying to draw themselves "just right." If time is running short, ask students take measurements now, but finish the drawing as a homework assignment.

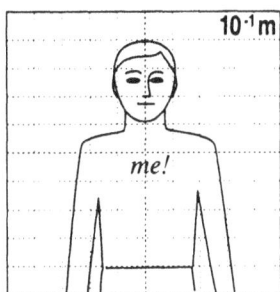

4-6.

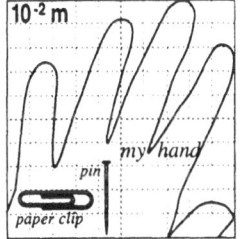 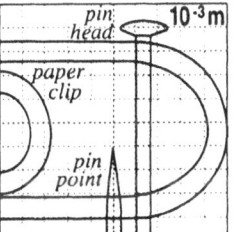 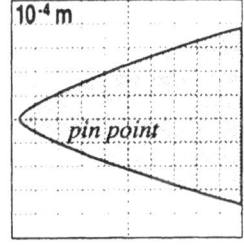

6. Students will sketch more realistic pin points if they ask themselves this question: If the pin rests across the tiny window so its point just meets the left edge, how much of the right side of the window is actually covered by the pin's widening taper?

MATERIALS

☐ A magnetic compass (optional).
☐ Book of Scale, pages 39-62, which yields 12 double-sided copies per student. These pages are "numbered" in small greyed type both front and back (center right on each page), in A, B, C, ... J, K, L order. Collate and paper-clip.
☐ Scissors.
☐ A high quality office stapler.
☐ A meter stick.
☐ A ruler or straightedge.
☐ A paper clip.
☐ A straight pin.
☐ A hand lens (optional).

NOTES: **Activity C1**

SCALE THE UNIVERSE (2)

Activity C2

1. Get a compass. Practice drawing arcs and circles on scratch paper to make sure it is in good working order....

... *Or,* follow these easy directions to make an alternative "*Circle Tool.*"

a. Stick about 11 cm of clear tape on a piece of wax paper. Cut around the tape's edge to make a long, skinny strip.

b. Poke a pinhole near one end. Make this hole large enough (by wiggling the pin) to pull the pinhead through it.

c. Stick a masking tape "handle" over the pin head.

d. Practice drawing arcs like this:

pencil in large pinhole

change radius (pin placement) for different circle sizes

2. Get your *Book of Scale,* and your *Log Scales* with plotted *Distance Tabs*.

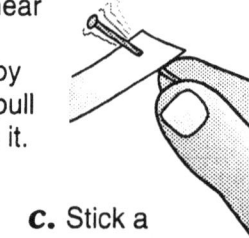

a. Practice by tracing over the "Proton" printed on the 10^{-16} m page. Notice that its radius is 870 "scale" attometers long, and that it is shown in the same corner as the tiny layout sketch on the *Log Scale*.

b. Turn to 10^{-15} m and draw the nucleus of a gold atom with a radius of 7 scale femtometers. Follow the tiny layout sketch to know where to draw it. (It's OK that other things are also on this page.)

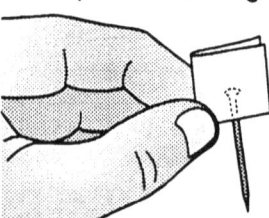

Label the size of each object as an equation.

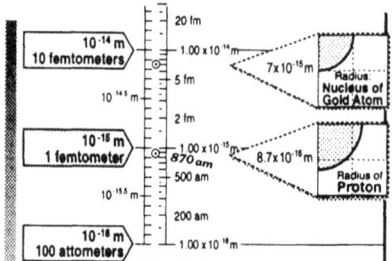

c. Draw each tab to scale, only once, on the page that fits best. (Not all pages will have drawings; some pages may have more than one.)

Drawing Guidelines:

1. Use a pencil with a good eraser. Be neat!
2. Place each drawing like its small tab layout.
3. Draw objects to scale. Use ruler and compass as needed. Sketch other shapes freehand. Be creative!
4. Use solid lines and shading for real objects. Use dashed lines to represent orbits and "concept" boundaries.
5. Label objects **and distance equations** in the units of that page (usually not meters).

OVERVIEW / OBJECTIVES

To redraw the thumbnail sketches on the Distance Tabs to scale in the Book of Scale. To practice drawing circles with a traditional compass, or improvised Circle Tool (optional) as preparation for making these scale drawings.

TIME: 1.5 hours.

NOTES / ANSWERS

1. These optional waxed-paper circle tools are easy to make and easier to control than traditional compass tools. If you already have enough traditional tools, check that they open and close smoothly, with enough friction to hold a fixed radius of any size, with pivots sharp enough to stay put. Cheaply made instruments are frustrating to use.

2. Selected Distance Tabs (the Proton, among others) are illustrated to serve as examples, and to inspire excellent work. Tabs drawn only once here, will be enlarged or reduced on adjacent pages in the next activity.

Student drawings should appear similar to the reductions detailed below.

Notice that scale drawings for distances to Sirius (10^{16} m), HD 70642 (10^{17} m), and beyond all use a sun symbol as the starting point (center left). At these vast astronomical distances, the Earth symbol (our home planet) and the Sun symbol (our home star) have the same location. Students should clearly understand that this Sun symbol and other star symbols do NOT represent the actual size of these objects. At these scales, these stars would be invisible if kept in accurate proportion.

MATERIALS

☐ Circle tools (optional). Improvise these with **clear tape**, **waxed paper**, **masking tape**, a **straight pin** and **scissors** as directed. Or use traditional compasses if they are in good working order.

☐ The Book of Scale constructed in the previous activity.

Answers 2.

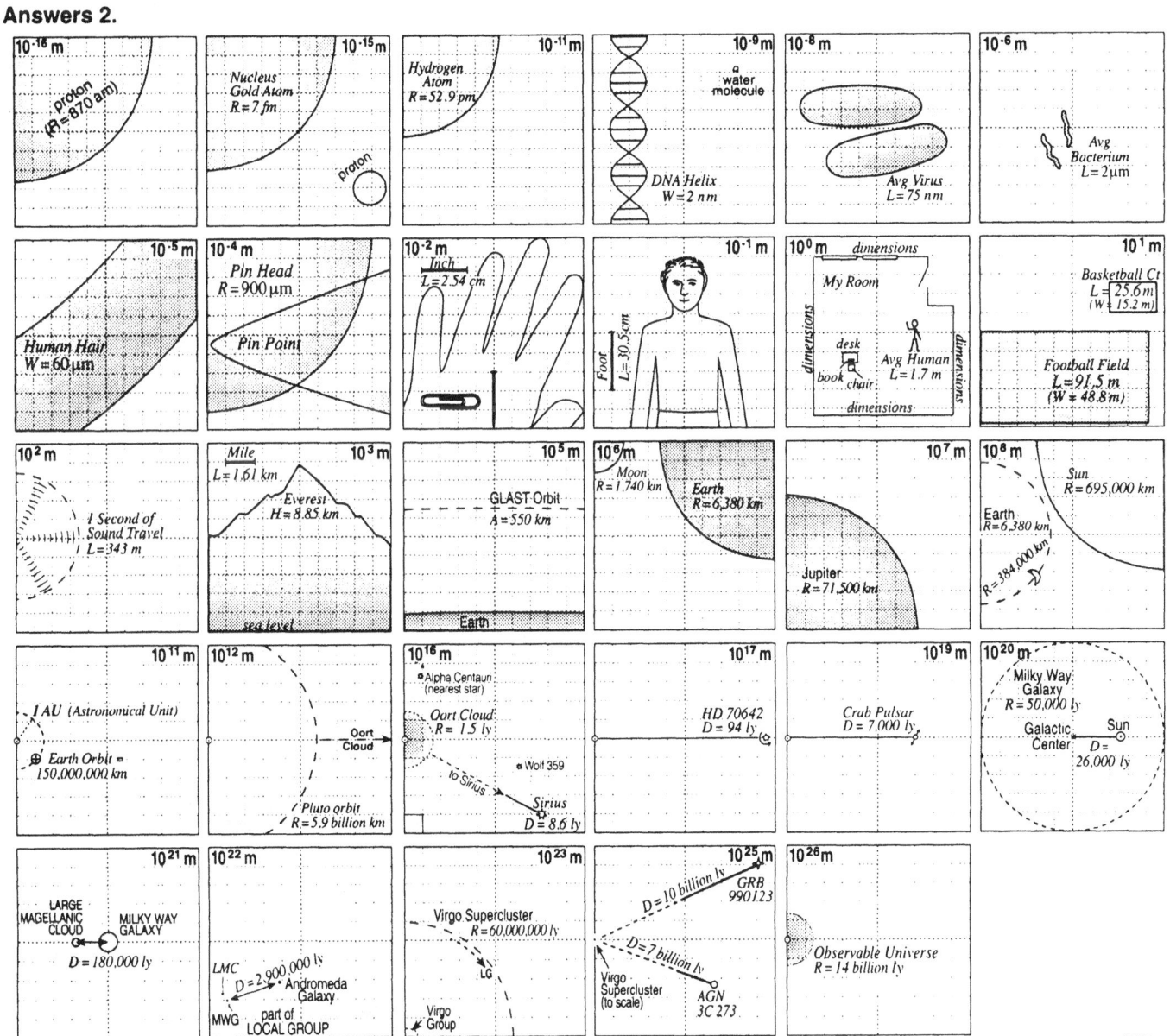

NOTES: **Activity C2**

SCALE THE UNIVERSE (3)

Activity C3

1. Make a tool that can draw really *big* circles:

  a. Cut a piece of string about a meter long. Tie an end around the sharp tip of a wood pencil (not a mechanical pencil).

  b. Tape the short end up the side of the pencil, so the loop can't slip off.

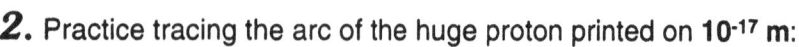

2. Practice tracing the arc of the huge proton printed on **10^{-17} m**:

  a. Measure out the proton's 870 am radius with 87 cm of string from the pencil. (There are 10 attometers per centimeter at this scale.)

  b. Tape this length of string to your table, and position your **Book of Scale** with the string crossing the "proton" diagonally, pencil point on the "proton surface." Draw the curve!

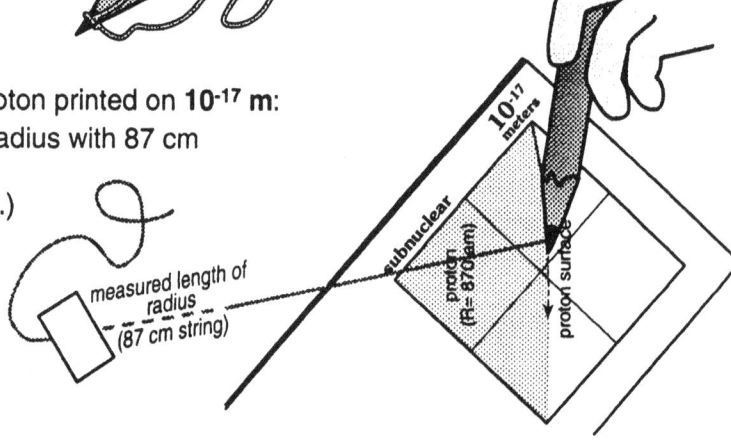

3. Now ENLARGE these objects 10X, just like the proton. Use these layouts to compose each drawing:

Draw objects on these pages.

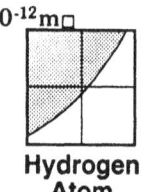
10^{-12} m
Hydrogen Atom

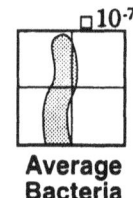
10^{-7} m
Average Bacteria

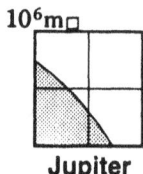
10^6 m
Jupiter

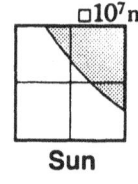
10^7 m
Sun

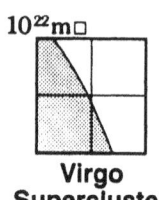
10^{22} m
Virgo Supercluster

4. REDUCE the following objects 10x, like the proton. Circle each scale in the list below when you have drawn and labeled that object on that grid. (Some are already illustrated for you, but may need labeling.)

Proton: ⟨10^{-17} m⟩ ⟨10^{-16} m⟩ ⟨10^{-15} m⟩ ⟨10^{-14} m⟩ ⟨10^{-13} m⟩
Nucleus of Gold Atom: 10^{-15} m, 10^{-14} m
Hydrogen Atom: 10^{-12} m, 10^{-11} m, 10^{-10} m, 10^{-9} m
Virus: 10^{-8} m, 10^{-7} m, 10^{-6} m
Bacteria: 10^{-7} m, 10^{-6} m, 10^{-5} m
Human Hair: 10^{-5} m, 10^{-4} m, 10^{-3} m
Pin: 10^{-4} m, 10^{-3} m, 10^{-2} m, 10^{-1} m
Paper Clip: 10^{-3} m, 10^{-2} m, 10^{-1} m
Average Human: 10^{-2} m, 10^{-1} m, 10^0 m, 10^1 m
Classroom: 10^0 m, 10^1 m

Football Field: 10^1 m, 10^2 m, 10^3 m
Mt. Everest: 10^3 m, 10^4 m *(Draw 100 km of Himalaya range that includes Mt. Everest. Use the bottom of the grid as sea level)*, 10^5 m
Moon: 10^6 m, 10^7 m, 10^8 m
Earth: 10^5 m, 10^6 m, 10^7 m, 10^8 m
Jupiter: 10^6 m, 10^7 m, 10^8 m
Moon's Orbit: 10^8 m, 10^9 m *(Center at left side of the grid, like larger drawing)*
Virgo Supercluster: 10^{22} m, 10^{23} m, 10^{24} m, 10^{25} m

5. Draw and label our solar system at five different orders of magnitude (from **10^{10} m** through **10^{14} m**)! Neatly show as much detail on each page as is practical.

AVERAGE RADIUS OF ORBIT FROM SUN

Mercury = 58 million km	Jupiter = 778 million km
Venus = 108 million km	Saturn = 1.43 billion km
Earth = 150 million km	Uranus = 2.87 billion km
Mars = 228 million km	Neptune = 4.50 billion km
Asteroid Belt = 414 million km	Pluto = 5.9 billion km

OVERVIEW / OBJECTIVES

To enlarge and reduce scale drawings by powers of ten on adjacent pages in the Book of Scale. To draw important features of the solar system to scale at 5 different orders of magnitude.

TIME: 1.5 hours.

NOTES / ANSWERS

1. Good eye-hand coordination helps a lot. Here are some ways to make this task easier: **(a)** First tape the pencil to the table so the point hangs over the edge. This leaves both hands free to tie the string. **(b)** Form a loop with a half knot first. Then snare the pencil point and pull tight. Finish by doubling the knot. **(c)** Dedicate several pencils to be "leashed," and leave them that way.

One more point: mechanical pencils also work. They are harder to tie up, but don't need to be resharpened if you dedicate several to be permanently leashed.

3. Students should draw the enlarged "average bacterium" freehand, 10X larger than it appears at 10^{-6} meters. The other enlargements should be drawn with the "leashed" pencil.

4-5. Most remaining circle drawings are within range of the traditional compass or waxed-paper Circle Tool. (The leashed pencil is used only in step 3.) Reductions smaller than a paper-punch should be drawn freehand, but with care.

MATERIALS

☐ Light string (kite string). Thread is too elastic to provide a constant radius for very large circles. And heavy cord forms big, clumsy knots.
☐ Scissors.
☐ Masking tape.
☐ A Circle Tool or traditional compass.
☐ Thread and scissors.

Answers 3-5. Answers continue on **notes C4.**

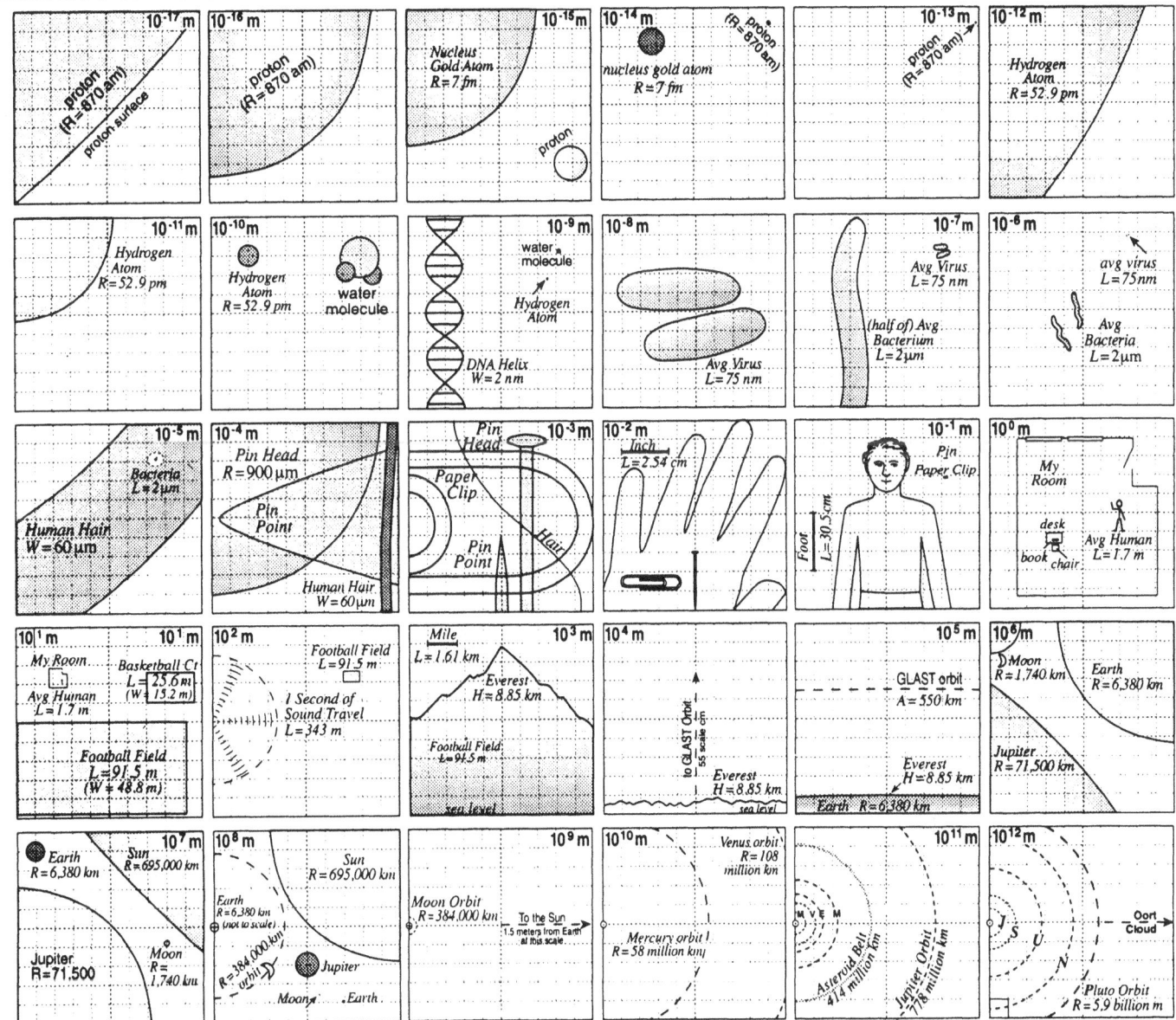

NOTES: **Activity C3**

PAGE 23

PROPORTIONAL THINKING

Activity C4

1. Cut out the *Human Scale Ruler*. Fold it in fourths along the solid black lines like an accordian, and attach a paper clip "marker" as shown.

  a. Open your *Book of Scale* to 10^0 m. Since each square on this grid is a giant step wide, how long is your classroom in giant steps?

  b. Assuming level, unobstructed walking, express these distances on a human scale:
  - length of a basketball court? (paper clip at 10^1 m.)
  - distance from sea level to the top of Mt. Everest? (paper clip at 10^3 m.)
  - distance to GLAST's orbit? (paper clip at 10^5 m.)
  - distance to the moon? (paper clip at 10^8 m.)
  - distance from Earth to Sun? (imagine paper clip at 10^{11} m.)

  c. Show that these distances jump by orders of magnitude: 12 minute walk; 2 hour walk; 2 day hike; 20 day hike; 200 day hike; 5.5 year hike; 55 year hike; 10 lifetimes.

2. Imagine that your body **expands** 8 orders of magnitude, while everything around you stays the same size. Fold over the *Human Scale Ruler* to the second row and move the paper clip to 10^4 m.

  a. As an exponential "+8" giant, you can now fit Mt. Everest (at 10^4) into an almost-invisible pore on your skin! How big is the Earth (at 10^7) in relation to your hand?

  b. Travel through the solar system and out to the stars. Describe how your surroundings look as a +8 giant.

3. Now imagine that your body **shrinks** by 9 orders of magnitude. (Fold your *ruler* to the third row.) Describe your amazing universe as a 10^{-9} midget:

  a. How big is a hydrogen atom (at 10^{-10}) at this scale?

  b. Travel through the microscopic world into the visible world. Describe how your surroundings look at your -9 size.

4. Explore the universe at other scales.

  a. What if you could walk all the way to the edge of the observable universe in under 3 days? (2 day hike @ 10^{26} m)

  b. What if a proton looked as big as a basketball? (hand width at 10^{-15} m)

  c. What if...

OVERVIEW / OBJECTIVES

To relate distances across vastly different orders of magnitude to human dimensions and how far we can walk. To develop an overview and kinesthetic sense of distance relationships in the Book of Scale.

TIME: 1 hour.

NOTES / ANSWERS

1a-b. Actual size: I can walk the length of my **n** meter classroom in **n** giant steps. Furthermore, I can walk these longer distances as follows:

BB court = 2.56 sq x 10 giant steps / sq = 26 giant steps
Mt. Everest = 8.86 sq x 12 min / sq = 106 minute walk
GLAST orbit = 5.5 sq x 2 day/ sq = 11 day hike
Moon = 3.84 sq x 5.5 yr/sq = 21 year hike
Sun = 1.5 sq x 100 lifetimes = 150 lifetimes of hiking

1c. 12 minutes x 10 = 120 minutes = 2 hours
2 hours x 10 = 20 hours = 2 day hike
2 days x 10 = 20 day hike
20 days x 10 = 200 day hike
200 days x 10 x 1 year / 365 days = 5.5 years
5.5 years x 10 = 55 years = 1 lifetime of walking
1 lifetime x 10 = 10 lifetimes.

2a-b. As a +8 giant (100 million times my actual size), Mt. Everest clogs a skin pore, while **Earth** fits into the palm of my hand like a **grapefruit**. Furthermore, I can walk ...

10^8 m: past the moon in 4 giant steps.
10^9 m: beyond Earth's gravity in 15 giant steps.
10^{11} m: from the Sun to Earth in 18 minutes.
10^{12} m: from the Sun to Pluto in under 12 hours.
10^{16} m: beyond our Sun's gravity in 8.25 years; to Sirus in about 48 years.
10^{21} m: across the Milky Way in 10,000 years.

3a-b. As a -9 giant (1 billionth actual size) a **hydrogen atom** fits in the palm of my hand like an **orange**. Furthermore....

10^{-13} m: A proton and the nucleus of a gold atom are still way too small to see.
10^{-9} m: I can across the width of DNA in 2 giant steps.
10^{-8} m: I can walk the length of a typical virus in 75 giant steps.
10^{-5} m: It takes 12 hours to cross the diameter of a human hair.
10^{-4} m: It takes 30 days to hike across a pinhead.
10^{-5} m: To travel 1 meter requires a lifetime of walking (55 years).

4. Students should fold their rulers to the bottom row, and pencil in various orders of magnitude ...

4a. 2 day hike @ 10^{26} m: If I can walk to the edge of the observable universe in under 3 days, then...

...I can walk to the center of our Virgo Super Cluster in a little over 7 minutes.
...I can walk to Andromeda in less than 30 giant steps.
...I can walk across our Milky Way Galaxy in 1 giant step.
...the distance between our Sun and Sirus fits inside a pore of my skin.

4b. hand width @ 10^{-15} m: If a proton fits into my hand like a basketball, then...

...it takes me over an hour to walk out of a hydrogen atom from its basketball center.
...it takes me 6 lifetimes to walk across the diameter of a human hair.

MATERIALS

☐ Scissors.
☐ A Human Scale Ruler, photocopy pg 64.
☐ A completed Book of Scale

Answers 3-5. Continued from **notes C3.**

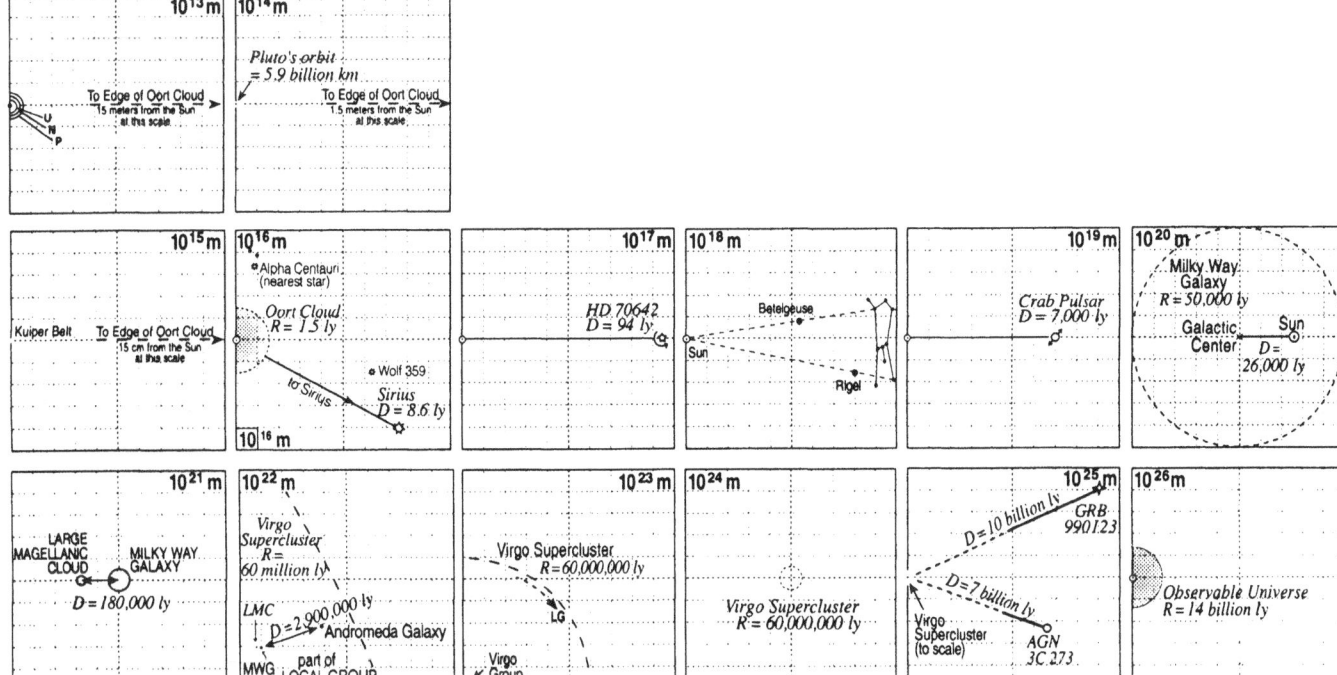

NOTES: **Activity C4**

ORDERING TIME

Activity C5

1. Get the *Time Tabs* page. Compare this actual ordering to what you decided as a class. For each category, write about what surprised you and what you learned.

2. Cut out the fastest time tab (.013 ys). Tape it under the fastest light clock (at 10^{-17} m in your *Book of Scale*), so its pointer is correctly positioned at 0.013 yactoseconds. Cut, position and tape all remaining *Time Tabs* in a similar manner to see how far light travels in given time spans.

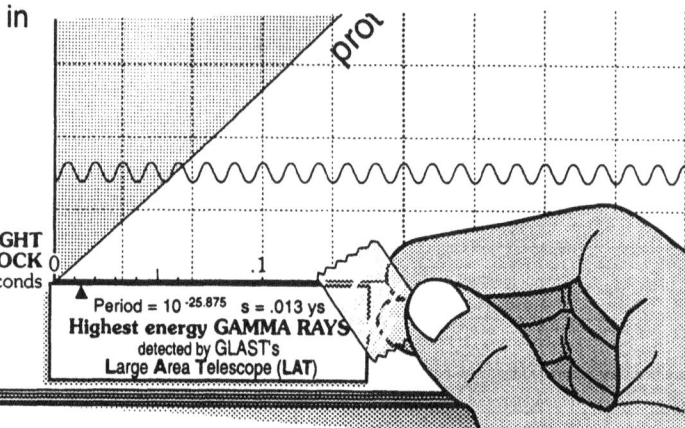

3. Summarize in a table how long it takes light from our Sun to reach each orbit:

TIME FOR LIGHT TO TRAVEL FROM SUN TO:			
Mercury	Mars	Saturn	Pluto
Venus	Asteroid Belt	Uranus	Kuiper belt *(outer limits)*
Earth	Jupiter	Neptune	Oort cloud *(outer limits)*

4. Answer the questions you find in your *Book of Scale* in the space provided. Begin at the outer limits of scientific knowledge (10^{26} m) and work towards the inner limits (10^{-17} m).

5. Going further...

a. Calculate **your age in seconds**. Write this in scientific notation. Take the log of your result to express your age as a base 10 exponent.

b. Compare **your age to a year on Pluto**. Do this two ways – by dividing numbers and subtracting exponents.

c. Add **new objects** to your Book of Scale: You might squeeze these into available white space, tape flip-up grid windows where needed, or construct a whole new book!

d. Construct a **base 2 Log Scale** with 2^n divisions. Measure, plot and label distances of interest.

e. Construct a **base 2 Book of Scale**.

OVERVIEW / OBJECTIVES

To qualitatively compare and sort time periods that range from nearly instantaneous to billions of years. To tape these time durations on "light clock" scales printed across the bottom of each grid in the Book of Scale. To develop a broad understanding of space, time and scale.

TIME: 3 hours (or more!)

INTRODUCTION

★ From distance (length, breadth and height), springs time, the fourth dimension. This introduction to time builds naturally on the introduction to distance presented earlier in activity B1 (page 15). And it follows the same sorting procedures. If your students have not yet sorted these distance tabs, please do those first.

Because time tabs are more abstract, you may want to read and decode meanings with your class as a first step. As a visual aid, photocopy the time tabs on page 63 for display on an overhead projector. Then present background information (below) for discussion and review.

Begin with **average** time durations, those closest to human experience; it's easiest to review and sort this category first. If your class is advanced, you might review all three categories at the same time, and even sort them together if you have 36 students. Otherwise, sort each category separately. Out of chaos comes order!

AVERAGE

Crab Pulsar: The remnant of a supernova explosion, witnessed by Chinese astronomers in 1054 to be as bright as the full moon! Today we see it through telescopes as a nebula with a rotating neutron star (pulsar) at its core. Like a lighthouse, it beams radio waves, visible light, X-rays and gamma-rays across our field of view once every 33 milliseconds. Over the next few million years it will lose energy and stop pulsing (rotating). Composed of protons and electrons crushed by gravity into an incredibly dense mass of neutrons, just one teaspoonful of a neutron star weighs as much as a mountain!

Long lasting GRBs / Brief GRBs: These are enormously powerful showers of gamma rays. Longer bursts can last about 90 seconds; others are as brief as just a few milliseconds. They are thought to be associated with the formation of black holes.

One Earth Rotation / One Earth Orbit: The Earth rotates once on its axis every 24 hours (1 day). It orbits once around the Sun in 365.25 days (1 year).

Sunlight travel to Earth: Light travels 1 astronomical unit (1 AU), across 150 million km of space, to reach Earth in just $8\frac{1}{3}$ minutes.

Period GLAST Orbit / GLAST Observation Window: The **G**amma **L**arge **A**rea **S**pace **T**elescope (GLAST) will be launched into a relatively low orbit, circling the Earth about 16 revolutions per day. This limits that amount of time that GLAST can "stare" at any particular place in the sky, without Earth getting in the line of view.

FAST

Electromagnetic periods: The time it takes electromagnetic radiation to travel one wavelength is called its period. Since all forms of light travel at constant speed (c = 3 x 10^8 m/s), it follows that the **Longest FCC Radio Waves** have periods many orders of magnitude longer than **Highest Energy Gamma Rays** with the shortest wavelengths.

Highest Energy Gamma Rays / Gamma Rays / Lowest Energy X-rays: GLAST has two kinds of radiation detectors. LAT (**L**arge **A**rea **T**elescope) will capture and track **Highest Energy Gamma Rays** up to 300 GigaelectronVolts (300 GeV). No previous space telescope has ever had the capacity to measure gamma rays this energetic. GBMs (**G**amma-ray **B**urst **M**onitors) will "see" radiation lower in the spectrum from **X-rays** to mid-energy **Gamma Rays**.

(Introduction continued on next page)

ANSWERS

3. Students should estimate these times from Light Clocks in their Books of Scale:

Mercury: 3 min	Asteroid Belt: 23 min	Neptune: 4.2 h
Venus: 6 min	Jupiter: 43 min	Pluto: 5.5 h
Earth: 8 min	Saturn: 1.3 h	outside Kuiper belt: 0.58 d
Mars: 13 min	Uranus: 2.7 h	outside Oort cloud: 1.6 yr

4. See pgs 30-31 for an answer key to all questions in the Book of Scale. Be sure students work from large to small scales. Doing good work on these questions might serve as an excellent take-home evaluation or assessment portfolio.

5a. This model answer is based on a student completing this problem on his or her 16th birthday:

$$\text{age}_{\text{years + days}} = 16 \text{ years} + 0 \text{ d}$$

$$\text{age}_{\text{days}} = 16 \text{ yr} \times 365.25 \text{ d/yr} = 5{,}844 \text{ d}$$

$$\text{age}_{\text{seconds}} = \frac{5{,}844 \text{ d}}{1} \times \frac{24 \text{ h}}{1 \text{ d}} \times \frac{60 \text{ min}}{1 \text{ h}} \times \frac{60 \text{ s}}{1 \text{ min}}$$

$$= 5.049 \times 10^8 \text{ s}$$

log 5.049×10^8 s = $10^{8.703}$ s = age of 16-year-old

5b. The Pluto time tab defines "1 year" as $10^{9.879}$ s, or 240 years. Comparing a Pluto year to the age of a 16 year old:

$$240 \text{ yr} / 16 \text{ yr} = 15 \text{ times shorter}$$

$$10^{9.879} \text{ s} / 10^{8.703} \text{ s} = 10^{1.176} = 15 \text{ times shorter}$$

5c. Individuals or group of students might develop their own Book(s) of Scale as a class project. If assigned as a cooperative exercise, different students might "adopt" different pages to fill with drawings.

5d-e. These log scales and books of scale will follow a slower pattern of exponential increase and decrease:

... $\frac{1}{8}$ m, $\frac{1}{4}$ m, $\frac{1}{2}$ m, 1 m, 2 m, 4 m, 8 m ...

MATERIALS

- ☐ TIME Tabs (sticky version), pg 63, 1 single-side copy.
- ☐ TIME Tabs (cutout version), pg 64, 1 single-side copy per student.
- ☐ Scissors.
- ☐ Clear tape.
- ☐ Book of Scale.
- ☐ A metric ruler.

NOTES: **Activity C5**

(Introduction continued from previous page)

SLOW

²/₅ Solar Mass Star / Sun / 15 Solar Mass Star / 40 Solar Mass Star: Large stars have more nuclear fuel to fuse than smaller stars, but their fusion reaction proceeds at a much faster rate. So this rule of thumb applies: the more massive the star, the faster it dies.

Precession of Earth's axis: Earth's axis wobbles, very slowly, like a spinning top. This is called precession. At our current position in the wobble, Earth's north axis points to Polaris. That's why all northern hemisphere stars appear to circle this "pole" star as the Earth turns.

Year on Pluto / Cosmic Year: "Year" means "once around." A year on Pluto (once around the sun) and a Cosmic Year (one turn of the Milky Way) are vastly different than an Earth year (once around the Sun).

Age of Observable Universe: Recall that the radius of the observable universe is almost 14 billion light years. So the CURRENT AGE of nothing we can "see" can be older than that. But what about LIFE SPANS that project into the future?

★ Electromagnetic energy radiates as electric charge vibrates. Electricity within an antenna oscillates millions of times per second to produce long, low-energy radio waves. Molecules vibrate trillions of times per second to produce shorter infrared waves of higher energy. Electric charge vibrates within atoms a thousand trillion cycles per second to produce ultraviolet waves, shorter than visible light. Atomic nuclei, quarks and other unstable particles vibrate at higher and higher and higher energies, producing a vast spectrum of gamma rays that GLAST will capture and "see" for the first time.

Summarize for your students with a simple diagram that surveys the entire spectrum.

gamma rays
X-rays
ultraviolet
visible
infrared
microwaves
radio

All of this radiation moves through the vacuum of space at 300,000,000 meters per second. Though different categories of "light" vibrate at vastly different frequencies, wavelengths and energies, it is fundamentally all of one piece. All light, all electromagnetic vibration, travels at a constant speed of 3×10^8 m/s, or about 1 foot per nanosecond.

Where do periods, listed on the fast Time Tabs, fit into this scheme? It is useful to examine this question in terms of unit analysis, a conceptual tool introduced in activity 2A. Recall that the period of electromagnetic radiation is defined as the time required for light to travel 1 wavelength:

$$\text{period} = \frac{\text{time}}{\text{wave}}$$

Frequency uses these same units, but turned upside down:

$$\text{frequency} = \frac{\text{waves}}{\text{time}}$$

Adding wavelength and speed to our analysis we have:

$$\text{wavelength} = \frac{\text{distance}}{\text{wave}}$$

$$\text{speed} = \frac{\text{distance}}{\text{time}}$$

Is there a way to multiply these units to help us understand new relationships? Here is a multiplication that goes nowhere useful, because the units don't cancel and simplify.

$$\frac{\text{waves}}{\text{time}} \times \frac{\text{distance}}{\text{time}} = \frac{\text{waves} \times \text{distance}}{\text{time}^2}$$

However, other multiplications reveal fundamental relationships. Challenge your students to discover these:

$$\frac{\text{waves}}{\text{time}} \times \frac{\text{distance}}{\text{wave}} = \frac{\text{distance}}{\text{time}}$$

$$\frac{\text{time}}{\text{waves}} \times \frac{\text{wave}}{\text{time}} = 1$$

$$\frac{\text{time}}{\text{waves}} \times \frac{\text{distance}}{\text{time}} = \frac{\text{distance}}{\text{wave}}$$

Thus: **frequency × wavelength = speed**
        period × frequency = 1
        period × speed = wavelength

So the next time you forget an equation in physics, just multiply units together and see where they lead.

Now to answer our original questions: How does period fit into the electromagnetic spectrum?

$$\text{period} = \frac{1}{\text{frequency}}$$

We see that period varies inversely with frequency, in the same direction as wavelength. This makes sense, because light traveling at a constant speed takes a longer time to traverse longer wavelength distances.

If you have displayed the diagram (left column) for your class, conclude by adding the words "and period" on the "wavelength" arrow.

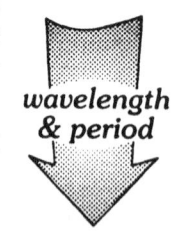

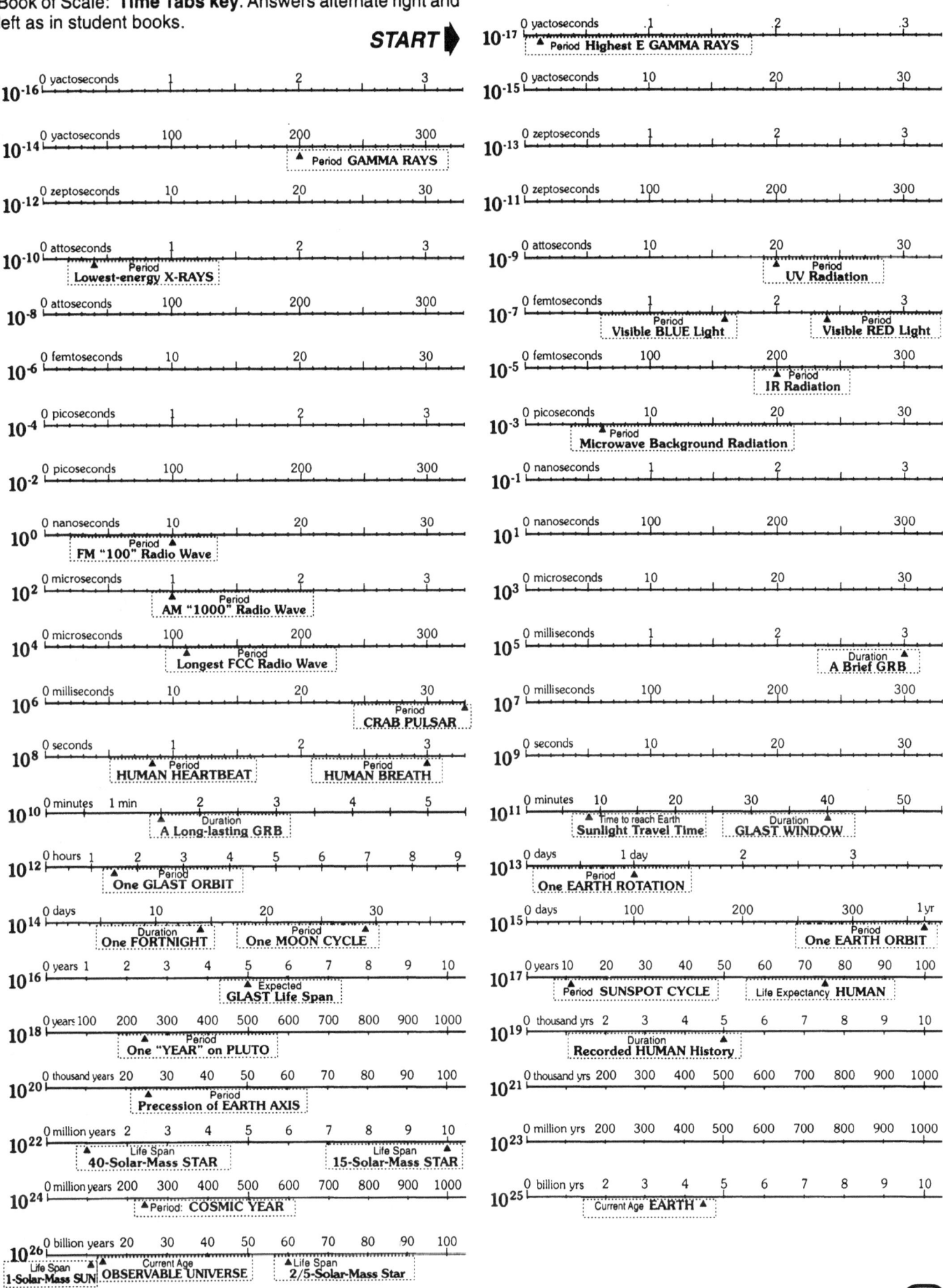

ANSWERS (continued from previous page) All *questions* found in the **Book of Scale** are answered in this key. Students should complete these questions in reverse order, starting from the largest scale in the back of the book, and working forwards toward smaller scales in the front.

10^{26} meters
a. 11 billion years ("Sun" tab)
b. 11 billion years/2 = 5.5 billion years
c. 4.5 billion years ("Earth" tab)
d. 5.5 billion years - 4.5 billion years
     = 1 billion years

10^{25} meters
Draw: 10^{25} m images 10X smaller @ 10^{26} m.
a. GRB = Gamma-Ray Burst;
   AGN = Active Galactic Nucleus
b. Short-term event: GRB 990123 is a flash of energy measured in seconds.
   Long-term object: AGN 3C 273 is the active heart of a galaxy with a life span of astronomical duration.
c. GRB energy output: a brief but spectacular burst of energy equivalent to a billion billion suns!
   AGN energy output: a continuous release of energy equivalent to a trillion suns. This is 6 orders of magnitude less intense than the GRB. But it lasts a whole lot longer!

10^{24} meters
life span Sun / cosmic year =
11 billion years / 240 million years =
11 billion years / 0.240 billion years =
  46 times around. Also,
$10^{17.540}$ m / $10^{15.879}$ = $10^{1.661}$ =
  46 cosmic years

10^{23} meters
Yes and no. A hypothetical Virgo civilization might eventually pick up our signals *if* these beings are still around in 60 million years when our signals, travelling at light speed, arrive.

10^{22} meters
No. A 40-solar-mass star lives about 10 million years. At death, its light would have traveled only about $\frac{1}{3}$ the distance to our Milky Way.

10^{21} meters
a. Nickel-sized Andromeda would be to the "right" and slightly "above" this Milky Way, at a scale distance of 2.9 windowlengths (29 cm). Each window length represents 1 million light years.
b. *Draw*, from the center of the grid, an arrow that rises to the right (about 3 cm of run with 1 cm of rise). Label it "To nickel-sized Andromeda: 29 cm at this scale."

10^{20} meters
age of Earth / cosmic year =
4.5 billion years / 240 million years =
4.5 billion years / 0.240 billion years =
     18.75 times around. Also,
$10^{17.153}$ m / $10^{15.879}$ = $10^{1.274}$ =
  18.79 cosmic years

10^{19} meters
GLAST can focus continuously on the Crab Pulsar for only about 40 minutes before it orbits "behind" the Earth.

10^{18} meters
a. No. Drawn to scale, the stars themselves would become invisible points.
b. At this scale, Sirus is less than 1 mm from the Sun, just inside the Sun-symbol circle.

10^{17} meters
This model answer is for the year 2007:
a. Star system HD 70642 is currently receiving Earth broadcasts transmitted 94 years ago, in the year 1913.
b. As of 1913, there were very likely no planets with life in star system HD 70642 that have evolved broadcasting technologies.

10^{16} meters
A dust particle 2 light years from the Sun is outside the Oort Cloud, and therefore beyond the pull of the Sun's gravity. It is likely orbiting another star.

10^{15} meters
a. The Kuiper Belt, with a radius of 15 billion km, represents 1.5 cm @ 10^{13} m, or a point with a 0.015 cm radius @ 10^{15} m. (This would be a barely visible dot with a diameter of .3 mm.)
b. Outer limit of the Oort Cloud: 15 cm from the Sun inside the Kuiper Belt dot.
c. radius Oort Cloud/ radius Kuiper Belt
   = 15 cm / 0.015 cm = 1000 times larger
➡ Comet Halley is a short period comet with an orbit likely confined within the Kuiper Belt. Comet Hale-Bopp, with a period 50 times longer than Halley, probably originated in the Oort Cloud, and spends most of its life orbiting the Sun in this more distant region of space.

10^{14} meters
After 1 complete orbit (about 27 days later), the moon isn't quite "full" again due to Earth's motion around the Sun. The moon needs 2 extra days to "catch up" to the new alignment between Earth and Sun. *(All 3 bodies in this diagram remain in the plane of the paper.)*

10^{13} meters
a. *Draw:* At center left, shade a "donut" beginning inside Pluto's orbit, just less than 0.5 cm from the Sun symbol, and extending 1.5 cm from Sun symbol.
b. Estimating time with the Light Clock, a gamma ray clears the Kuiper Belt in just under 0.6 days or 14 hours.
c. *Draw:* @10^{14} m, draw a tiny semi-circle 1.5 mm from the Sun position (center left). Label it "Outer Limit of Kuiper Belt."

10^{12} meters
Draw: Shade and label a Kuiper-Belt "donut" wedge starting 4.5 cm from the Sun (center left), and extending 15 cm into the space at the right of the grid.
a. *Draw:* Extend and calibrate the Light Clock beyond the grid to the right.
b. The Kuiper-Belt reaches to about 14 hours on the Light Clock, or 0.6 days as estimated in the previous problem.

10^{11} meters
Astronomical unit (1 AU): the average distance between Earth and Sun.
a. 1 AU = 150 million km
   = 150 billion m = 8.33 light minutes
b. Jupiter in AUs:
   778 million km / 150 million km = 5.2 AU

10^{10} meters
Draw: a segment of Earth's orbit in the open space to the right of the grid. This arc has a radius that extends 15 cm from the Sun location (center left).
a. *Draw:* An arrow between Earth orbit and the Sun labeled "1 AU."
b. *Draw:* The Sun to scale as a tiny circle with a radius of 0.7 mm (diameter = 1.4 mm).

10^9 meters
a. *Draw:* A point about 1.5 million km (4 moon orbits or 1.5 squares), to the right of the center-left Earth symbol. Label it "LaGrange Point."
b. This transiting asteroid orbits the Sun. At 2 million km from Earth, it is 0.5 million miles beyond the LaGrange Point, where the Sun's gravitational pull exceeds the Earth's.

10^8 meters
➡ The moon symbol is not to scale.
a. (As given)
b. $(1.8 \times 10^{10}$ m/sec$)(60$ sec/hr$) =$
   108×10^{10} m/hr = 1.08×10^{12} m/hr
   (Check answer @ 10^{12} m)
c. $(1.08 \times 10^{12}$ m/hr$)(24$ hr/day$) =$
   25.9×10^{12} m/day = 2.59×10^{13} m/day
   (Check answer @ 10^{13} m)
d. $(2.59 \times 10^{13}$ m/day$)(365.25$ day/yr$) =$
   946×10^{13} m/yr = 9.46×10^{15} m/yr
   (Check answer @ 10^{15} m)

10^7 meters
a. about 10 Earths fit across Jupiter
b. about 10 Jupiters fit across the Sun
c. about 100 Earths fit across the Sun

Book of Scale (answer key, continued)

d. About 1,000 Earths fit inside Jupiter.
e. About 1,000 Jupiters fit inside the Sun.
f. About 1,000,000 Earths fit in the Sun.

10^6 meters
a. 6,380 km / 1,740 km =
   3.67 moons fit across Earth
b. $3.67^3 \approx 50$ moons

10^5 meters
If the scale height of 1 grid square represents 100 km, the height of the full window represents 1,000 km. Since the moon's real orbit is 384,000 km above Earth, you'd need to increase this grid's height by 384 windows (38.4 meters).

10^4 meters
a. wavelength = 3.33 scale cm
   = 33.3 actual meters
b. 1 wave passes by in 111 μs.
   3 waves pass by in 333 μs.
   9 waves pass by in 1 ms.
   9,000 waves pass by in 1 s.
c. frequency = 9,000 c/s
d. distance moved in 1 second
   = 300 thousand km
e. speed = c = 300 thousand km/sec
   = 3×10^5 mm/sec
   = 3×10^8 m/sec

10^3 meters
a. Everest is roughly 2 orders of magnitude taller than a football field is long.
b. 8,860 m / 91.5 m = 97 times
or, $10^{3.947}$ m / $10^{1.961}$ m = $10^{1.986}$ = 97 times
c. log 97 = 1.99
Mt. Everest is precisely 1.99 OM's taller.

10^2 meters
a. 1.6 km/mile × 1,000 m/km
   × 1 s/343 m = 4.66 s ≈ 5 s/mile
b. When you see lightning, count the seconds until you hear thunder. Divide this number by 5 to estimate the number of miles.

10^1 meters
a. A track star runs the 100 meter dash in 10 s, or 10^{10} ns. (Turn 10 pages or 10 orders of magnitude between clocks.)
b. A radio wave "runs" 100 m in 333 ns.
c. track star time / radio wave time =
   1×10^{10} ns / 333 ns =
   0.003×10^{10} times faster =
   3×10^7 or 30 million times faster

10^0 meters
You are listening to FM radio. An AM station broadcasts waves that are a thousand or so meters long, while an FM station broadcasts waves that are only meters long. Moving the radio right or left makes a difference in reception relative to shorter FM waves of comparable order of magnitude.

10^{-1} meters
a. Light travels 1 foot in 1 ns (1 ft/ns).
b. Students should measure their eye-thumb distance with arm extended. This equals roughly 2 ns on the light clock, depending on arm length.

10^{-2} meters
Light travels 1 inch / 8.4 ps. Thus:
   1 inch / 8.4 ps × 10^{12} ps/s
   × 1 m / 39.37 inches = 3×10^8 m/s

10^{-3} meters
Microwave background radiation has a wavelength about twice as long as the diameter of paper clip wire.

10^{-4} meters
a. *Draw* red blood cell:
   0.84 mm diameter dot
b. *Draw* human egg:
   1 cm diameter circle
c. *Draw* human sperm:
   2.5 mm long super-thin wavy line

10^{-5} meters
a. *Label:* 2 femtoseconds as the "period of yellow light" on Light Clocks at 10^{-5} m (almost at zero), 10^{-6} m (near zero), and 10^{-7} m (between "blue" and "red" tabs).
b. *period:* IR radiation takes 100 times longer to travel 1 wavelength compared to yellow light.
   energy: IR radiation has 100 times less energy compared to yellow light. *(Energy and wavelength are inversely proportional.)*

10^{-6} meters
➻ *Label* the Light Clock: blue at 1.6 fs and red at 2.4 fs (left & right of yellow.)
➻ Visible light washes over tiny virus, as ocean waves over small rocks, but bounces off larger bacteria into our eyes.

10^{-7} meters
a. Blue's period is shorter (than red's);
b. Blue's frequency is higher;
c. Blue's wavelength is shorter:
d. Blue's energy is higher.

10^{-8} meters
length average virus = 750 Å
length average bacterium = 20,000 Å
20,000 Å / 750 Å = 27 times longer

10^{-9} meters
a. 2 OM's more energetic than yellow light: This is UV radiation with a period of 20 attoseconds, and a wavelength of 6 nanometers.
b. 2 OM's less energetic than yellow light: This is IR radiation with a period of 200 femtoseconds, and a wavelength of 60 micrometers.

10^{-10} meters
diameter of hydrogen atom = 1 Å
width of human hair = 600,000 Å
So 600,000 hydrogen atoms fit side by side across a human hair.

10^{-11} meters
10^0 s = 1 second = 1 s
10^{-3} s = 1 millisecond = 1 ms
10^{-6} s = 1 microsecond = 1 μs
10^{-9} s = 1 nanosecond = 1 ns
10^{-12} s = 1 picosecond = 1 ps
10^{-15} s = 1 femtosecond = 1 fs
10^{-18} s = 1 attosecond = 1 as
10^{-21} s = 1 zeptosecond = 1 zs
10^{-24} s = 1 yactosecond = 1 ys

10^{-12} meters
a. pin prick diam = 1.74 femtometers
b. 1 mm represents 100 femtometers at this scale
100 / 1.74 = 57.5 pin pricks in 1 mm
c. Since a hydrogen atom's radius is 52.9 pm, the beachball's radius is 52.9 cm. So its diameter is 1.058 meters.
d. If a hydrogen atom's nucleus were a barely visible pin prick, the whole atom would be as big as a 1-meter beachball!

10^{-13} meters
Calibrate: major divisions on the Light Clock as 1,000, 2,000 and 3,000 yactoseconds.
Draw: a wave with a period of 200 ys on both grids. At 10^{-14} m, it is 6 squares long; at 10^{-13} m, only 0.6 squares long.

10^{-14} meters
LAT = Large Area Telescope
GBM = Gamma-ray Burst Monitor
GLAST = Gamma Large Area Space Telescope

10^{-15} meters
Gold's heavier nucleus is made of many protons and neutrons *(197 on average)*.

10^{-16} meters
1 Ym = 1 yottameter = 10^{24} m
1 Zm = 1 zetameter = 10^{21} m
1 Em = 1 exameter = 10^{18} m
1 Pm = 1 petameter = 10^{15} m
1 Tm = 1 terameter = 10^{12} m
1 Gm = 1 gigameter = 10^9 m
1 Mm = 1 megameter = 10^6 m
1 km = 1 kilometer = 10^3 m
1 m = 1 meter = 10^0 m
1 mm = 1 millimeter = 10^{-3} m
1 μm = 1 micrometer = 10^{-6} m
1 nm = 1 nanometer = 10^{-9} m
1 pm = 1 picometer = 10^{-12} m
1 fm = 1 femtometer = 10^{-15} m
1 am = 1 attometer = 10^{-18} m

10^{-17} meters
a. 7.5 waves in 1/10 yactosecond
b. 75 waves in 1 yactosecond
c. 75×10^{24} waves in 1 second
d. 75×10^{24} waves / second
e. $1 / 75 \times 10^{24}$ =
   0.0133×10^{-24} seconds/wave =
   0.0133 yactoseconds per wave =
   period written on the time tab

NOTES: **Activity C5** (continued)

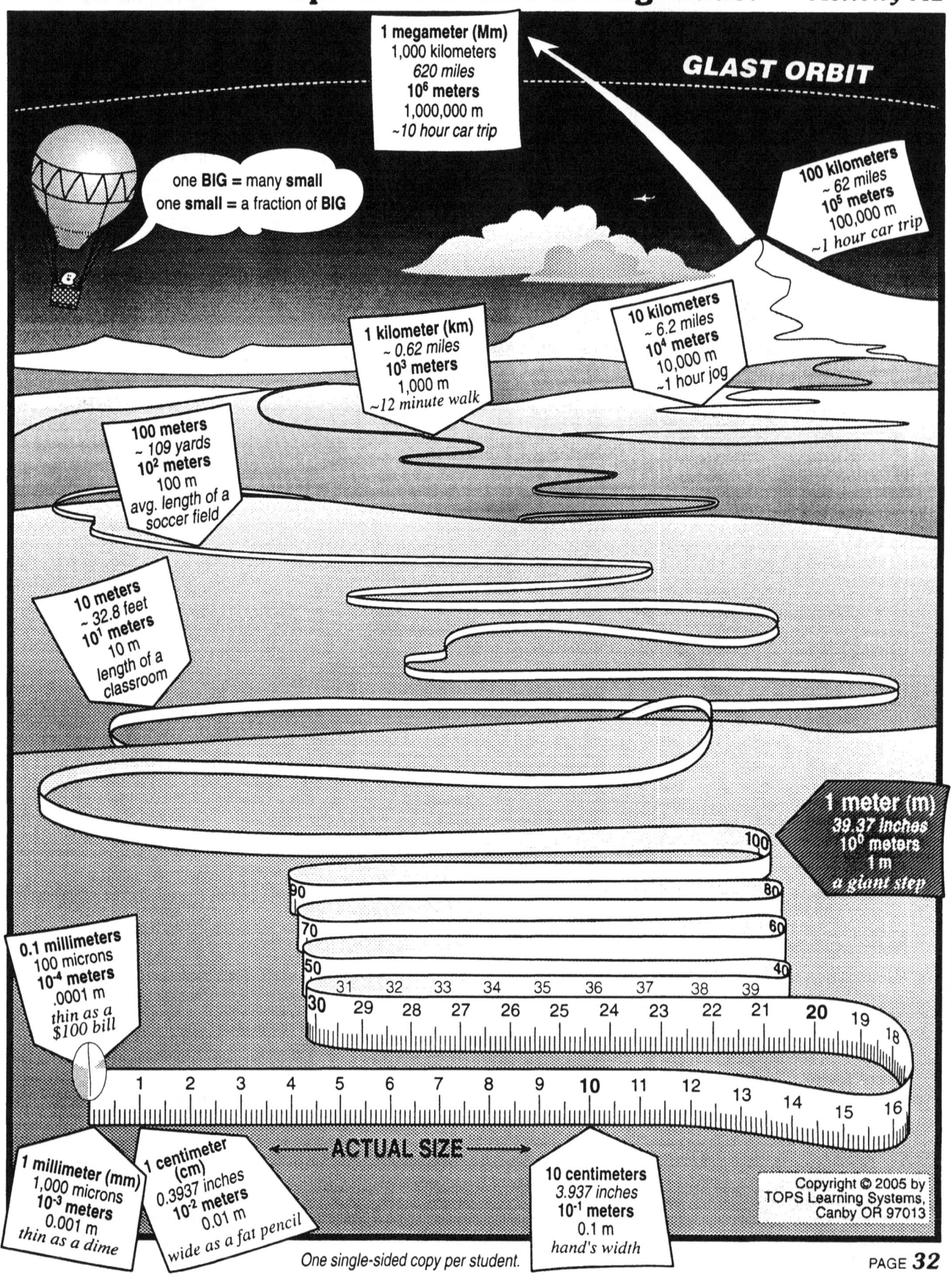

Individual/Small Group Activity:
ORDERING DISTANCE ("Sticky Tab" Version)

1. Cut out four sections along the heavier dashed lines.
2. Stick masking tape along grey bars, half off the paper.
3. Chop the "Human Scale" strip into separate "sticky tabs" on lighter dashes.
4. Distribute tabs to any number of students as detailed on page 15.
5. Tape final sequence along a table, desk or on wall.
6. Repeat for the other three categories.

human scale

Altitude: GLAST Orbit — Gamma-ray Large Area Space Telescope

One FOOT

Radius of HYDROGEN ATOM

Length of Average BACTERIUM

Width of DNA Helix

Radius of PROTON (nucleus of hydrogen atom)

Radius: NUCLEUS of Gold Atom

Length of Average VIRUS

human scale (middle strip)

Radius of PIN HEAD

One MILE

Height of Average HUMAN

Length of FOOTBALL FIELD

Width of HUMAN HAIR

Height of Mount EVEREST — Earth's tallest mountain

One INCH

Distance SOUND travels in 1 SECOND (thunder follows lightning one mile away by 5 seconds)

Length of BASKETBALL COURT

solar system and nearby stars

Radius of JUPITER

Distance to SIRIUS — brightest star in our night sky

Radius of EARTH

Distance to HD 70642 — a Sun-like star with a Jupiter-like planet

Radius of our MOON

Radius of OORT CLOUD — objects within this limit still orbit our Sun

Radius of EARTH ORBIT — 1 Astronomical Unit (A.U.)

Radius of PLUTO'S ORBIT

Radius of MOON'S ORBIT — Average distance from Earth

Radius of SUN

astronomical scale

Radius of VIRGO SUPERCLUSTER (our Local Group of galaxies rotates near outer edge)

Distance to CRAB PULSAR — Spinning neutron star in constellation Orion

Radius of Observable UNIVERSE

Distance to GALACTIC CENTER of our Milky Way

Distance to GRB 990123 — a temporary Gamma Ray Burst; equal to the energy of a billion billion Suns

Distance to LMC — Large Magellanic Cloud; a dwarf satellite galaxy of our Milky Way

MILKY WAY GALAXY — from center to edge

Distance to ANDROMEDA — largest galaxy in our Local Group

Distance to AGN 3C 273 — Active Galactic Nucleus; sustained energy of a trillion Suns

Copyright © 2005 by TOPS Learning Systems, Canby OR 97013 — Copy 1 single-sided page per class. — PAGE 33

DISTANCE TABS
("Cutout Tab" Version)

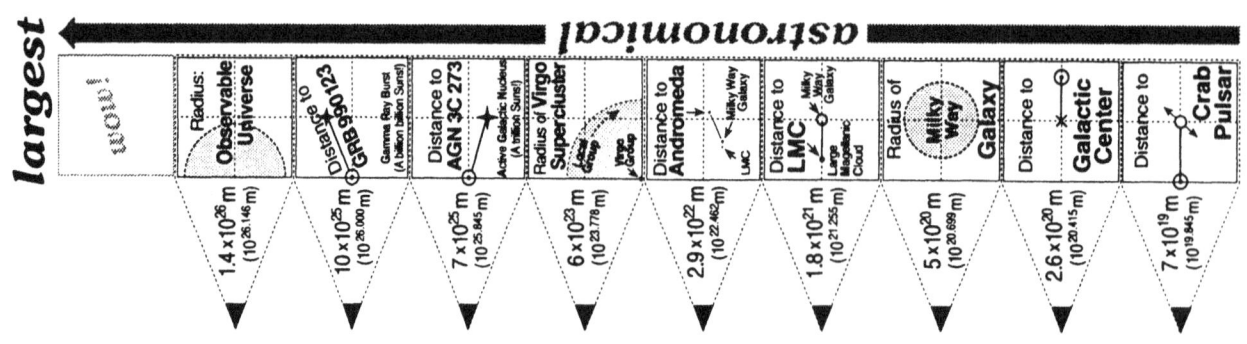

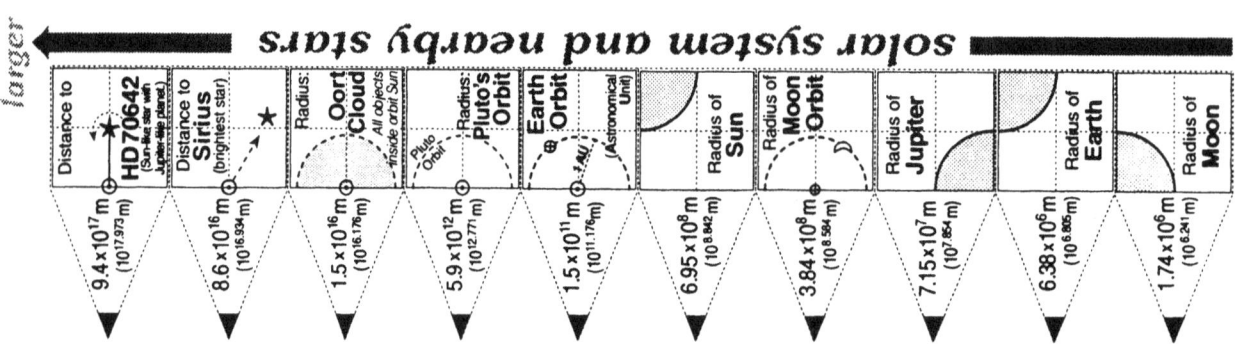

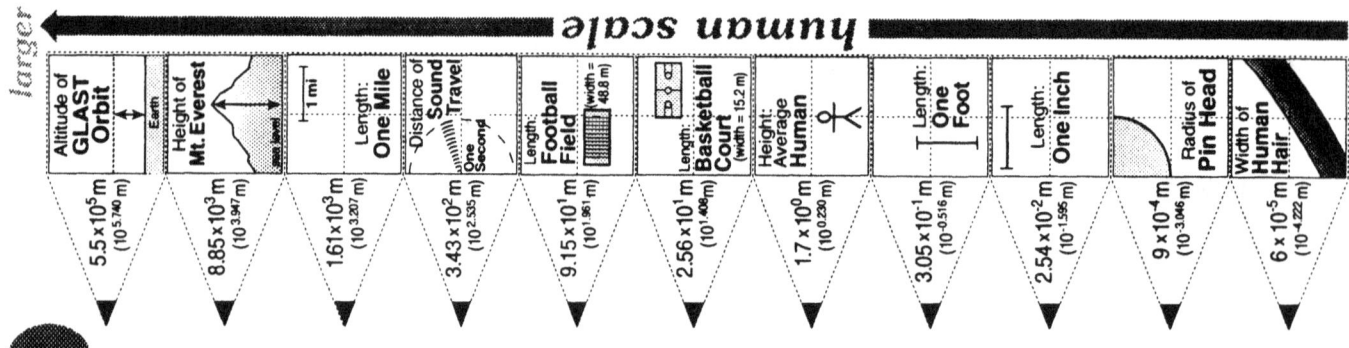

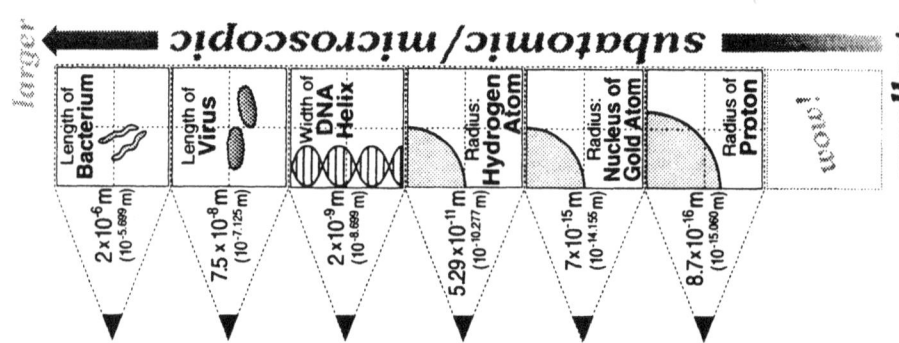

PAGE **34** — Copy 1 single-sided page per student. — Copyright © 2005 by TOPS Learning Systems, Canby OR 97013

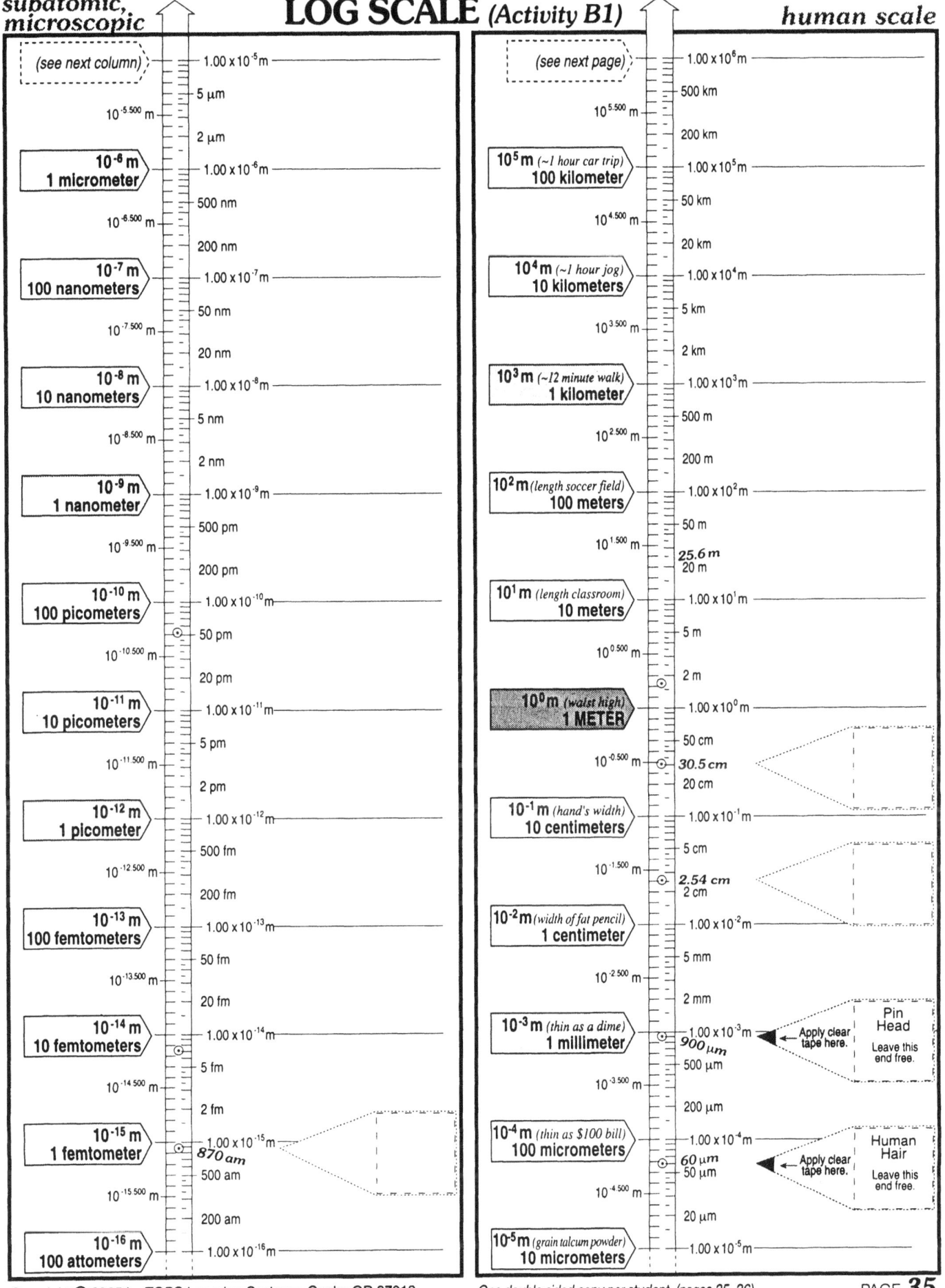

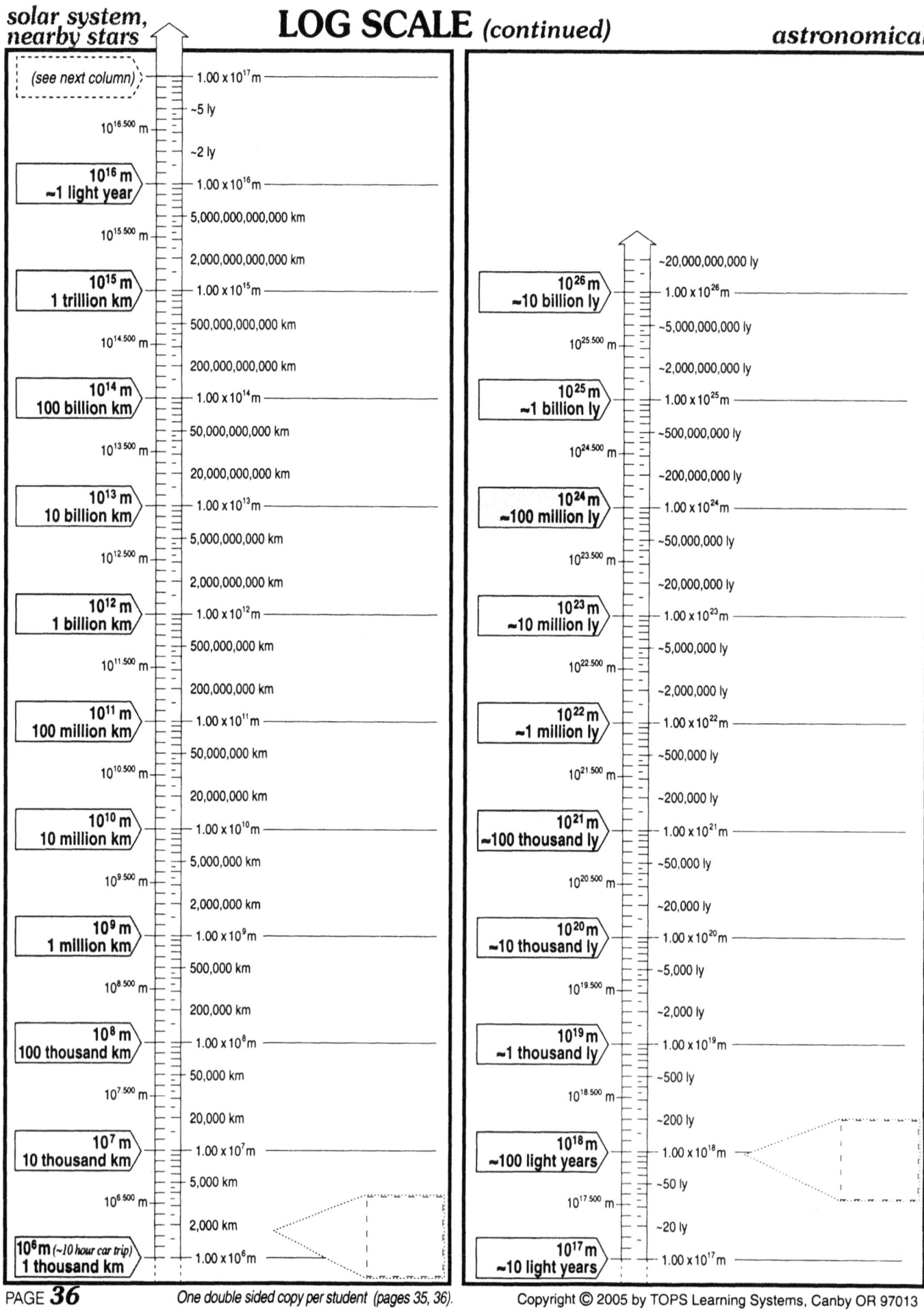

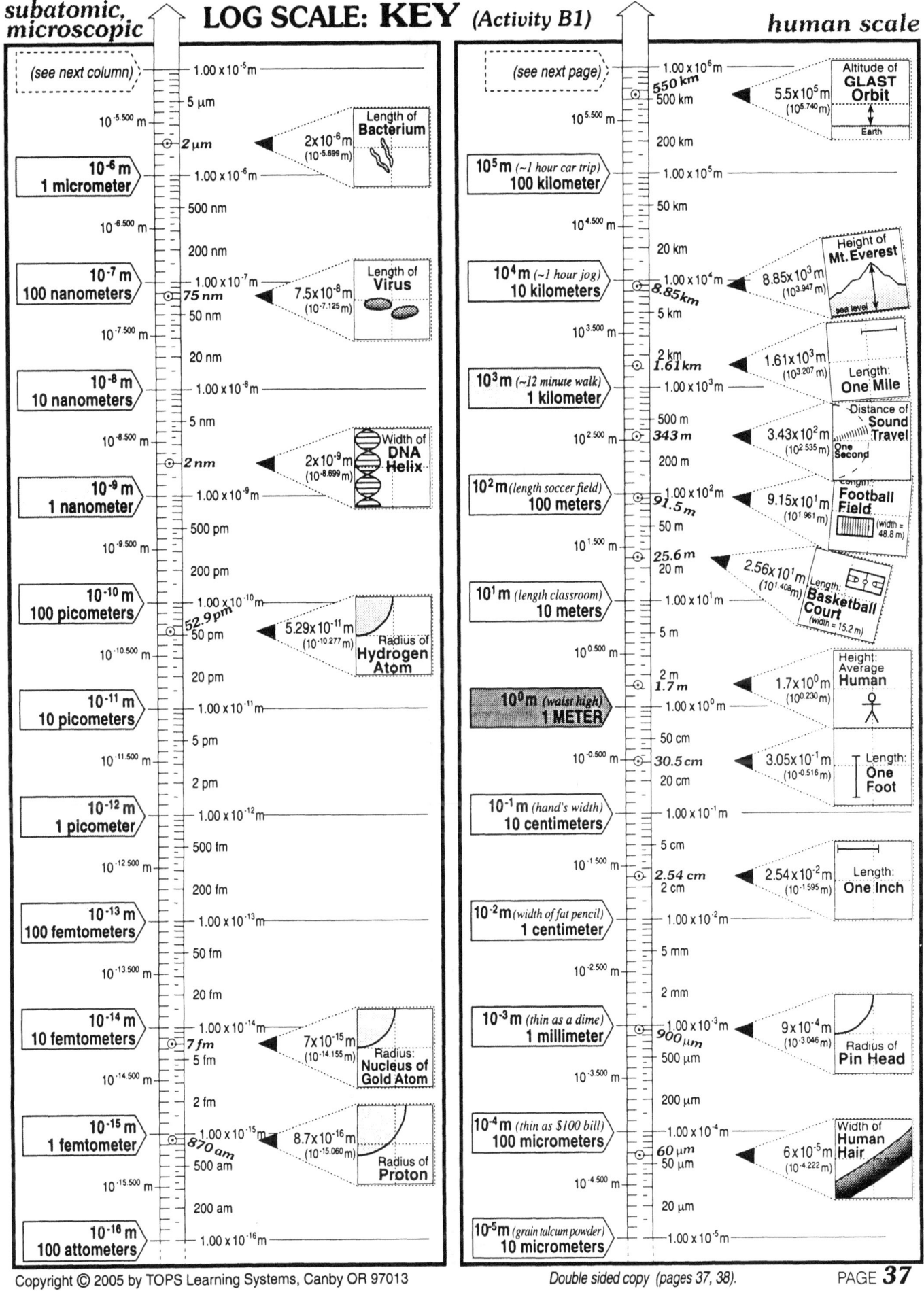

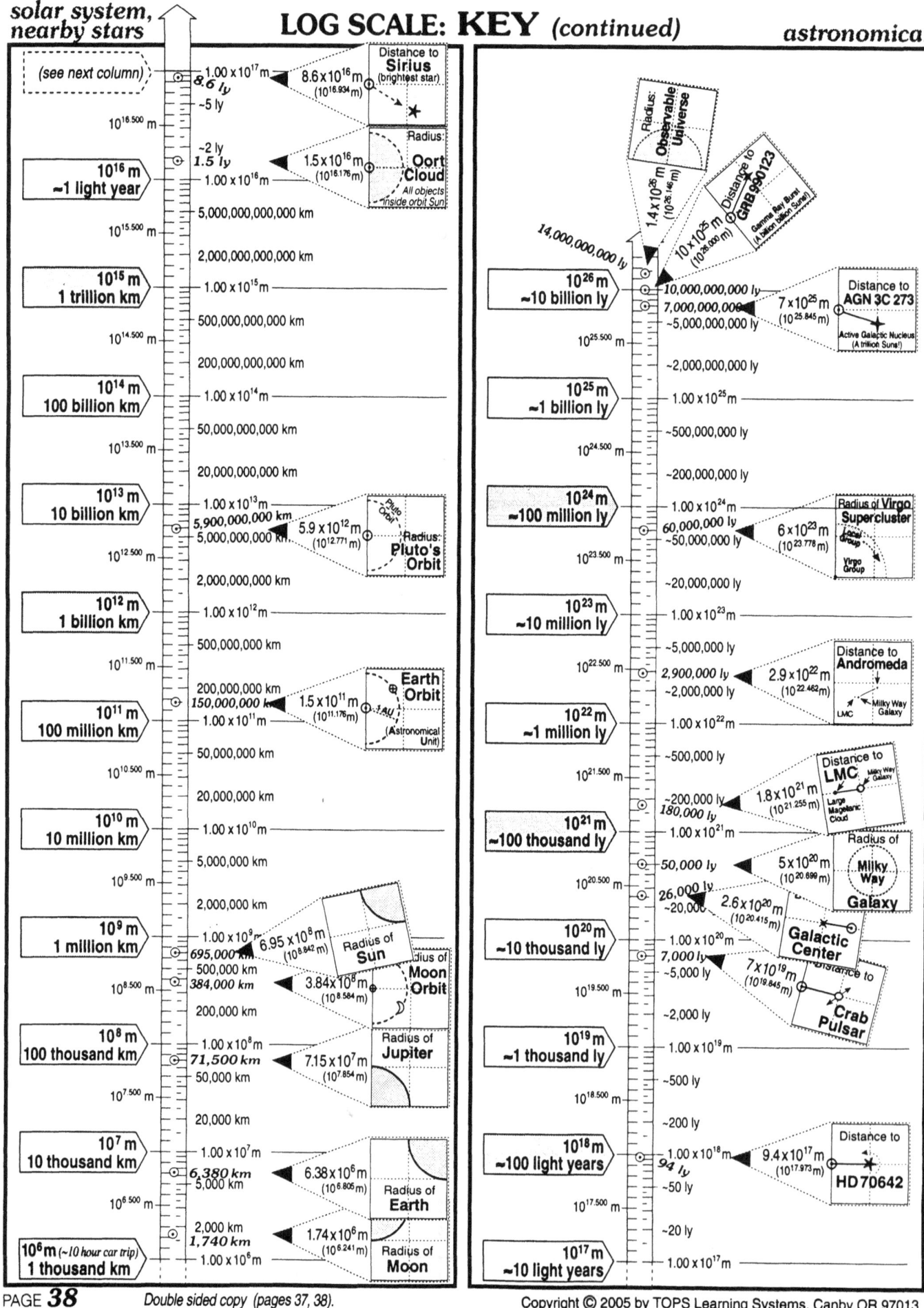

Activities C1-C5

BOOK OF SCALE

All Things Big *and* Small

SMALL to LARGE by powers of 10 →

name:

FRONT

---

subnuclear

1 cm represents **10 attometers** or 0.01 fermi

10^{-17} **meters**

➥ A wavetrain from the highest energy photon that GLAST can capture is shown crossing this grid. How many waves whiz by in...

a. in 1/10 yactosecond?

b. in 1 yactosecond?

c. in 1 second?

d. Frequency = waves / second. What is the frequency of this gamma ray?

e. Show that frequency and period are inversely related. You should get the same period as this time tab!

frequency = 1 / period.

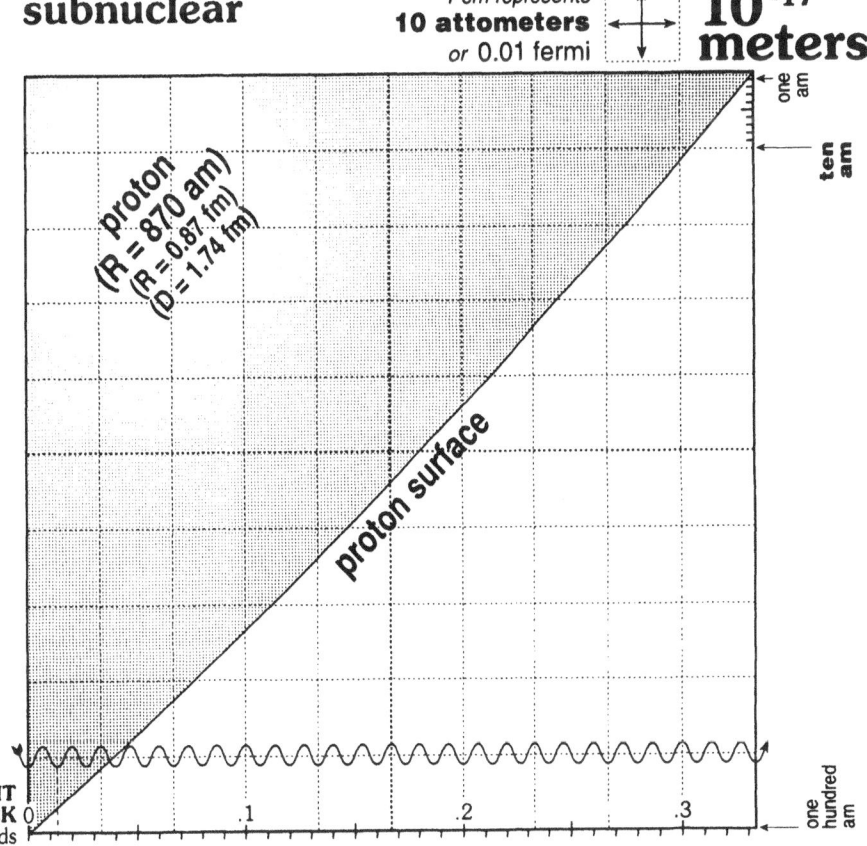

proton (R = 870 am) (R = 0.87 fm) (D = 1.74 fm)

proton surface

LIGHT CLOCK 0 .1 .2 .3
yactoseconds

one am
ten am
one hundred am

The front of this book investigates the SMALLEST OBJECTS! SHORTEST DISTANCES! BRIEFEST TIME PERIODS!

Copy/collate 1 set of 12 double-sided pages per student.

This booklet is a component of the TOPS book SCALE THE UNIVERSE.
Copyright © 2005 by TOPS Learning Systems, Canby OR 97013

A: back

10^{-16} meters

1 cm represents **100 attometers** or 0.1 fermi

proton or neutron

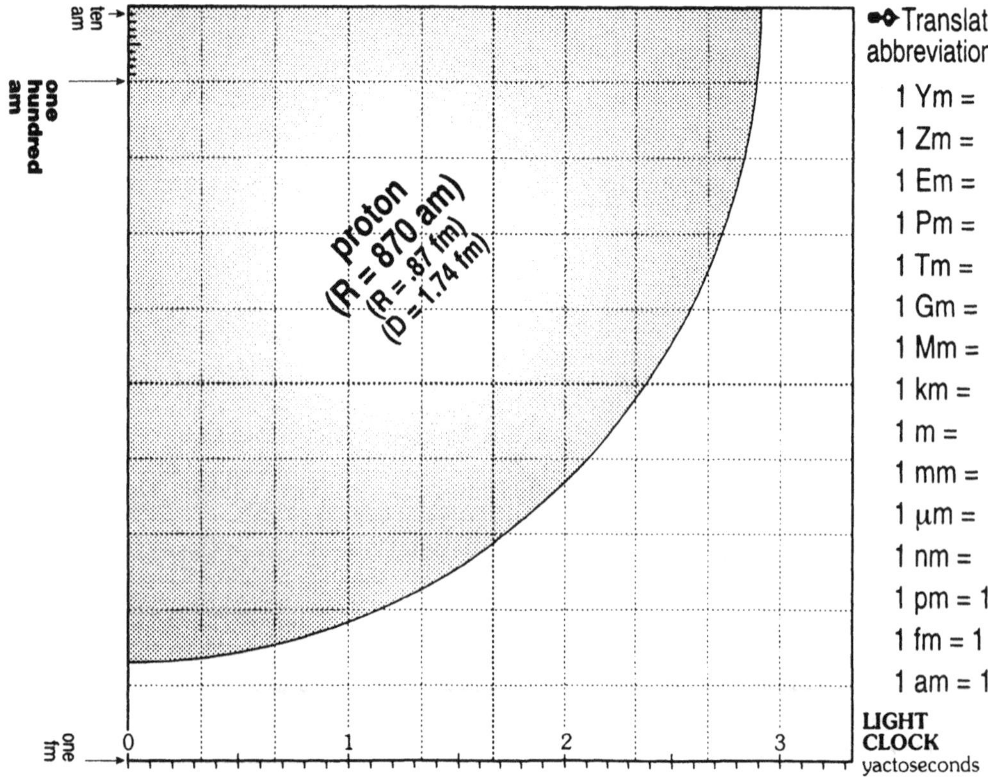

proton
(R = 870 am)
(R = .87 fm)
(D = 1.74 fm)

ten am →
one hundred am →
one fm →

0   1   2   3

➡ Translate this official list of metric abbreviations, working up from the bottom:

1 Ym =
1 Zm =
1 Em =
1 Pm =
1 Tm =
1 Gm =
1 Mm =
1 km =
1 m =
1 mm =
1 μm =
1 nm =
1 pm = 1 picometer = 10^{-12} m
1 fm = 1 femtometer = 10^{-15} m
1 am = 1 attometer = 10^{-18} m

LIGHT CLOCK
yactoseconds

→ Why is the 1-proton nucleus of a hydrogen atom so much smaller than the nucleus of a gold atom?

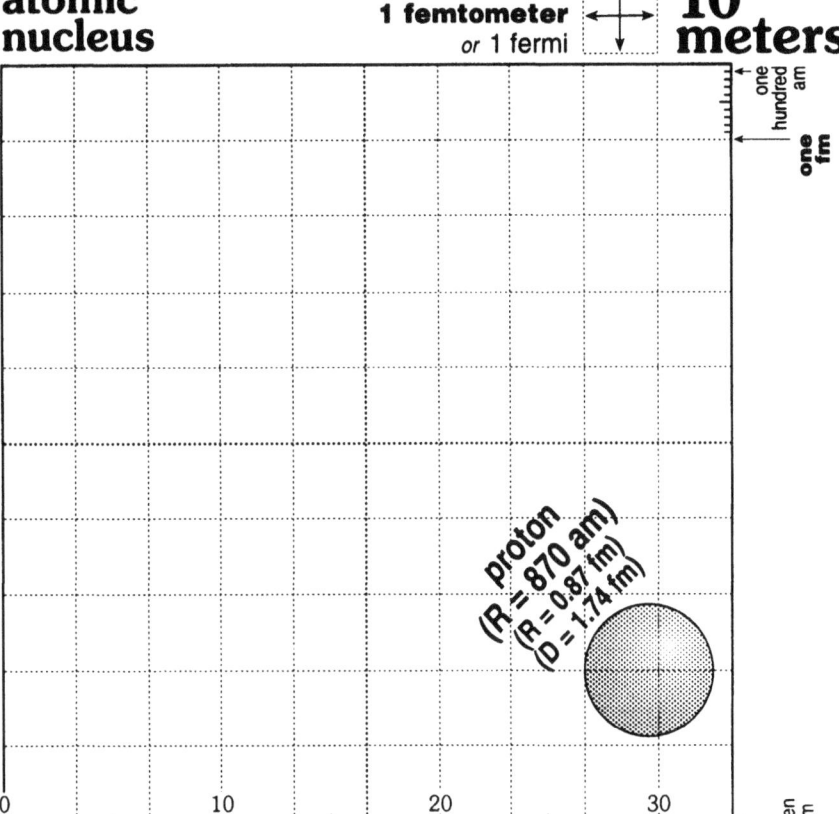

→ Label the Light Clock calibrations at the bottom of this grid in equivalent yactoseconds.

→ Draw a gamma wave with a period of 200 yactoseconds on both the 10^{-13} m and 10^{-14} m grids.

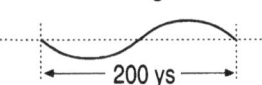

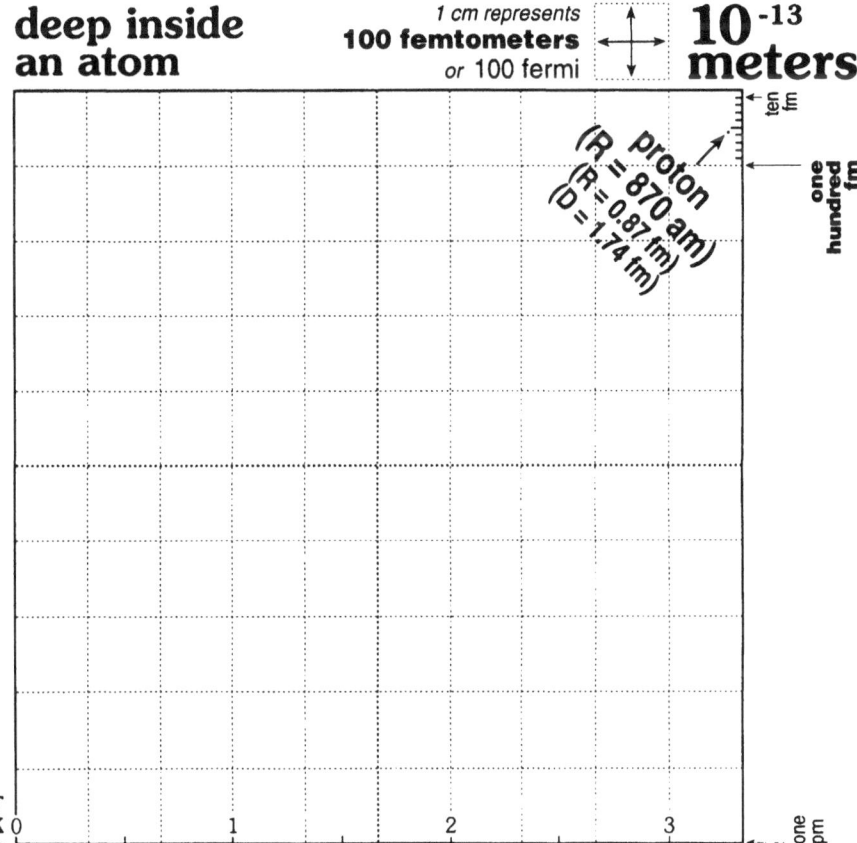

10^{-14} meters

1 cm represents **10 femtometers** *or 10 fermi*

near the nucleus

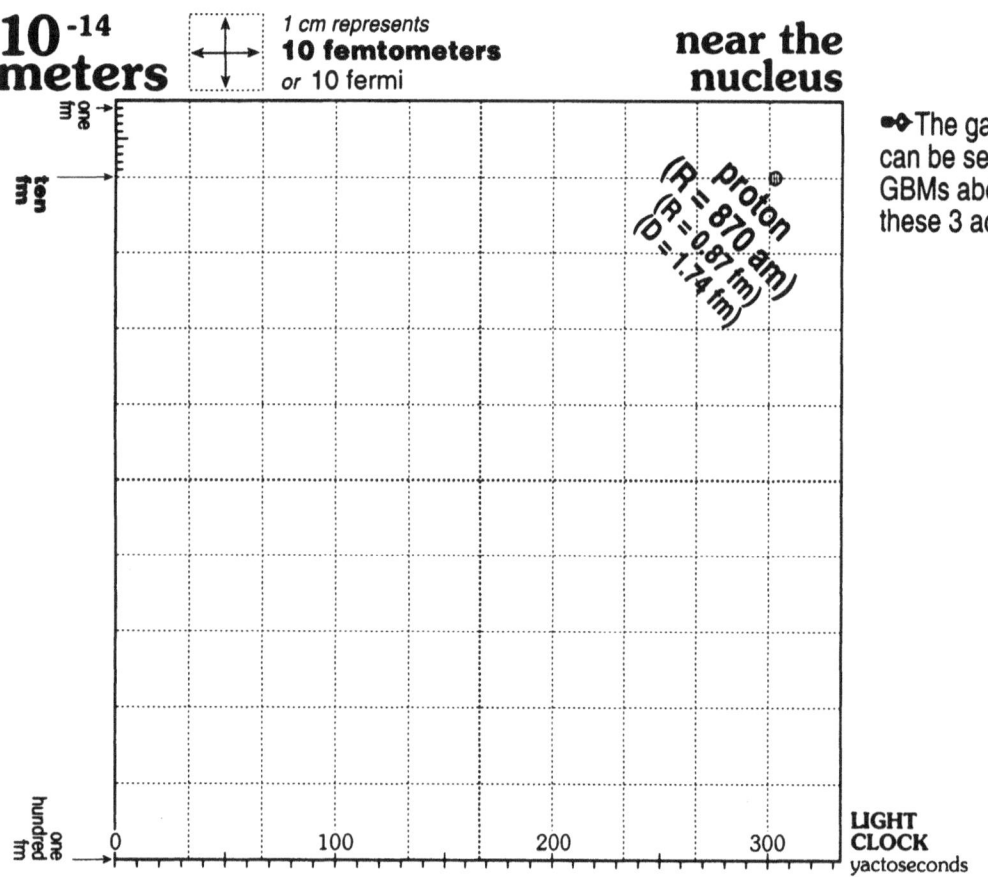

→ The gamma waves you've drawn can be seen by both LAT and the GBMs aboard GLAST. Translate these 3 acronyms:

10^{-12} meters

1 cm represents **1 picometer** *or 1,000 fermi*

still inside an atom

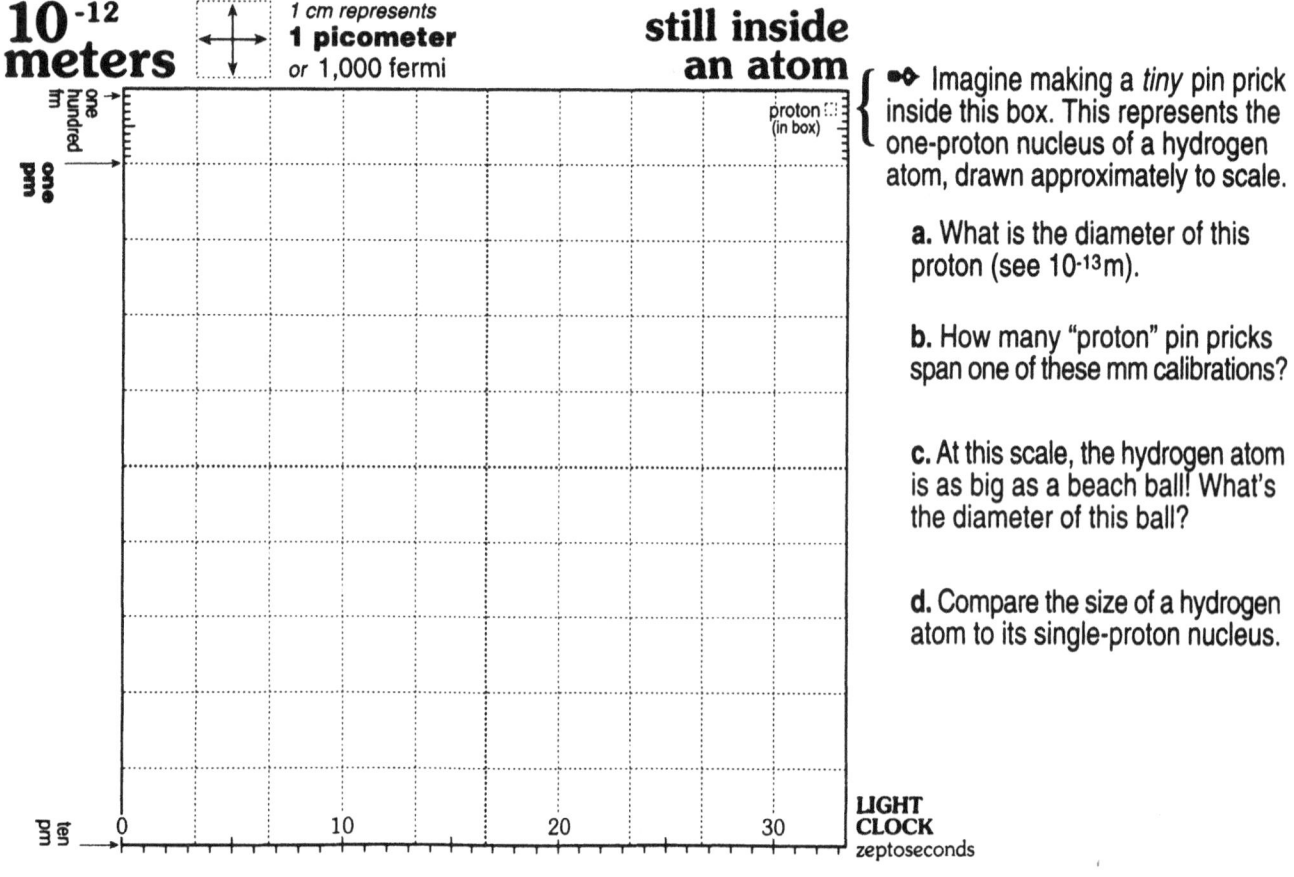

→ Imagine making a *tiny* pin prick inside this box. This represents the one-proton nucleus of a hydrogen atom, drawn approximately to scale.

a. What is the diameter of this proton (see 10^{-13} m).

b. How many "proton" pin pricks span one of these mm calibrations?

c. At this scale, the hydrogen atom is as big as a beach ball! What's the diameter of this ball?

d. Compare the size of a hydrogen atom to its single-proton nucleus.

atoms

1 cm represents **10 picometers** *or 0.1 angstrom* = 10^{-11} **meters**

➺ Complete this metric time table:

10^0 s = 1 second = 1 s

10^{-3} s = 1 millisecond = 1 ms

10^{-6} s =

10^{-9} s =

10^{-12} s =

10^{-15} s =

10^{-18} s =

10^{-21} s =

10^{-24} s =

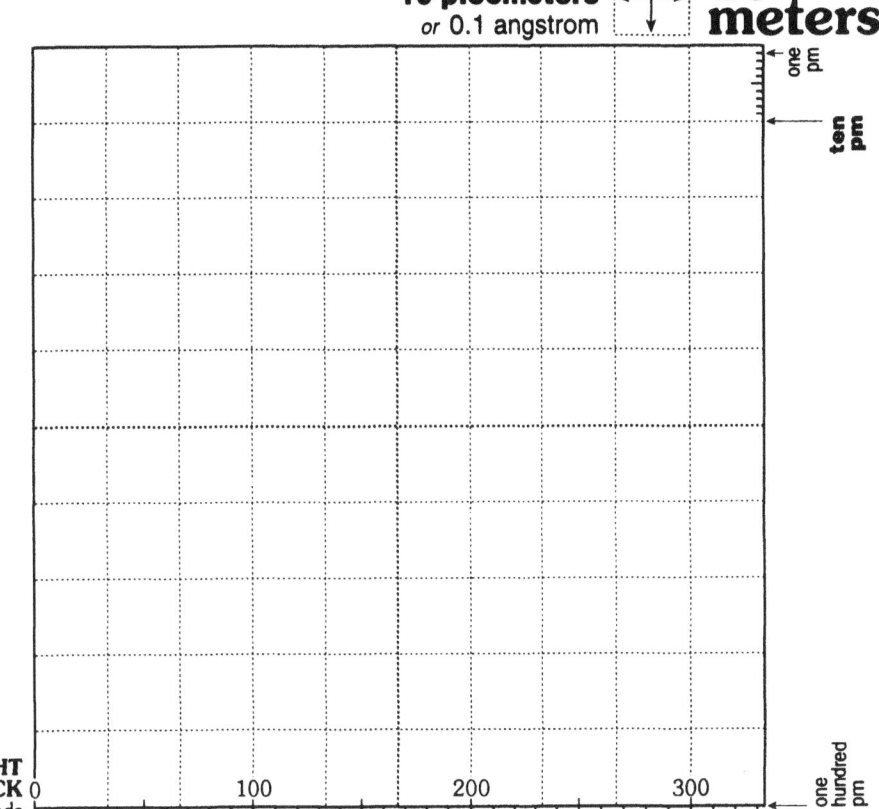

LIGHT CLOCK zeptoseconds: 0, 100, 200, 300

DNA

1 cm represents **1 nanometer** *or 10 angstroms* = 10^{-9} **meters**

➺ Fully describe electromagnetic radiation that is...

a. Two orders of magnitude *more energetic* than yellow light.

b. Two orders of magnitude *less energetic* than yellow light.

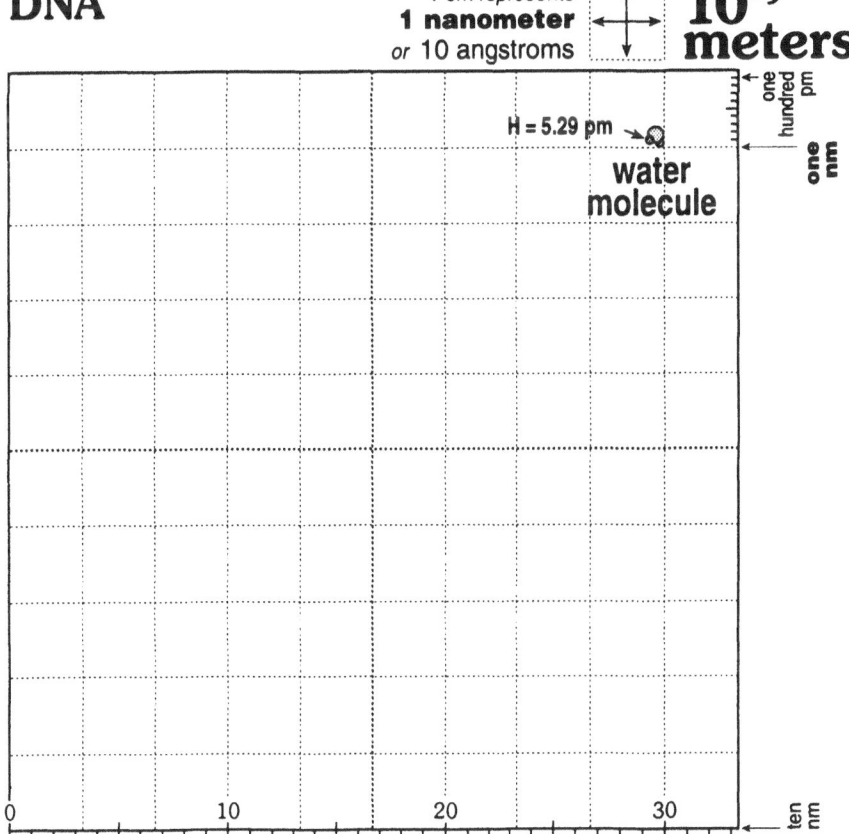

H = 5.29 pm — water molecule

LIGHT CLOCK attoseconds: 0, 10, 20, 30

10^{-10} meters — simple molecules

1 cm represents **100 picometers** or 1 angstrom

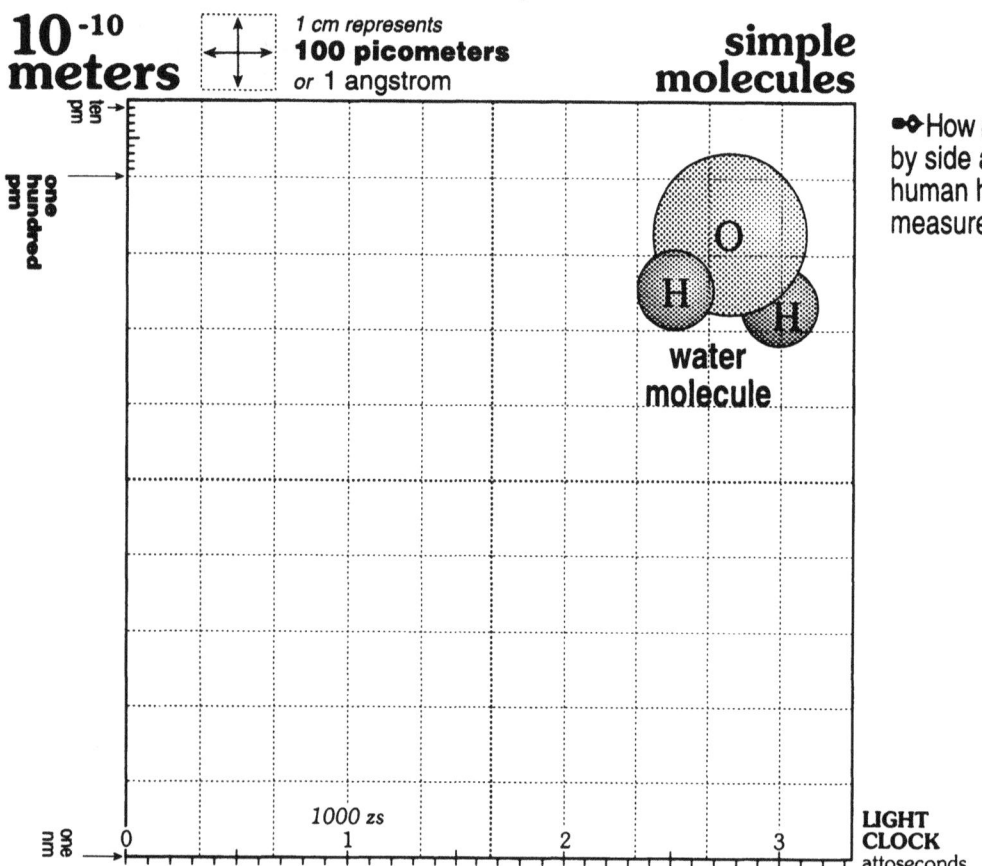

➡️ How many hydrogen atoms fit side by side across the diameter of a human hair? (First determine each measure in angstroms.)

LIGHT CLOCK attoseconds

10^{-8} meters — virus

1 cm represents **10 nanometers** or 100 angstroms

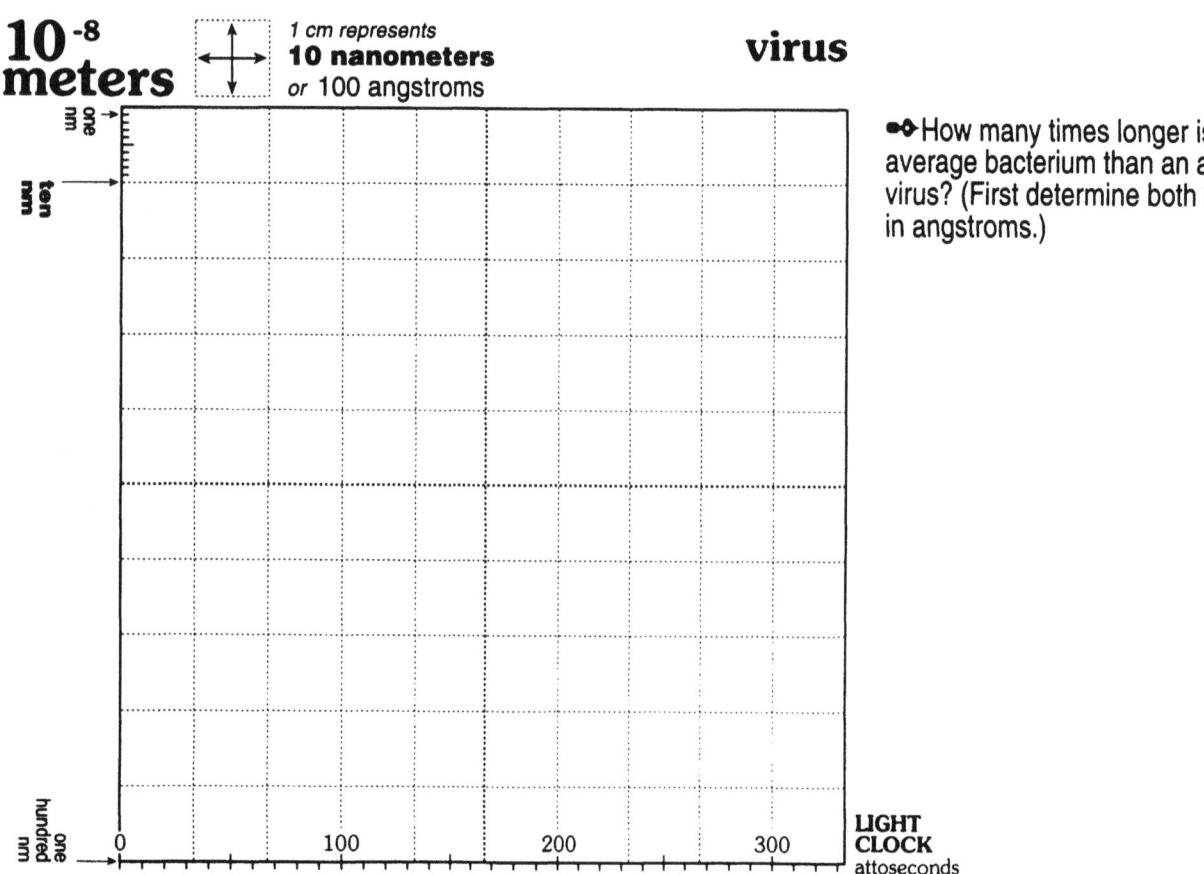

➡️ How many times longer is an average bacterium than an average virus? (First determine both lengths in angstroms.)

LIGHT CLOCK attoseconds

↦ Compare blue and red light in terms of...

  a. period:

  b. frequency:

  c. wavelength:

  d. energy:

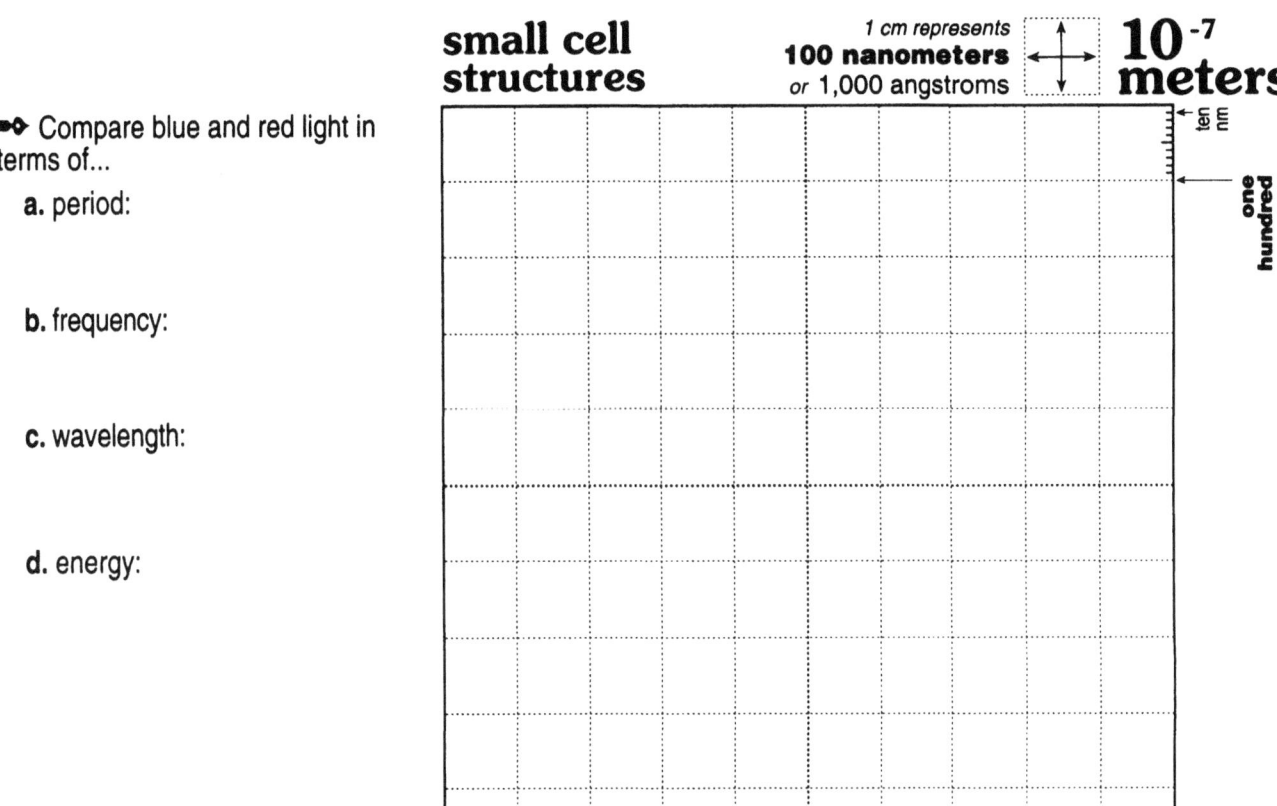

↦ A certain wavelength of yellow light has a period of 2 femtoseconds.

  a. Label this period on Light Clocks under these grids: 10^{-5} meters, 10^{-6} meters, 10^{-7} meters.

  b. Compare IR radiation, defined on the time tab below, to yellow light with respect to...

   period:

   energy:

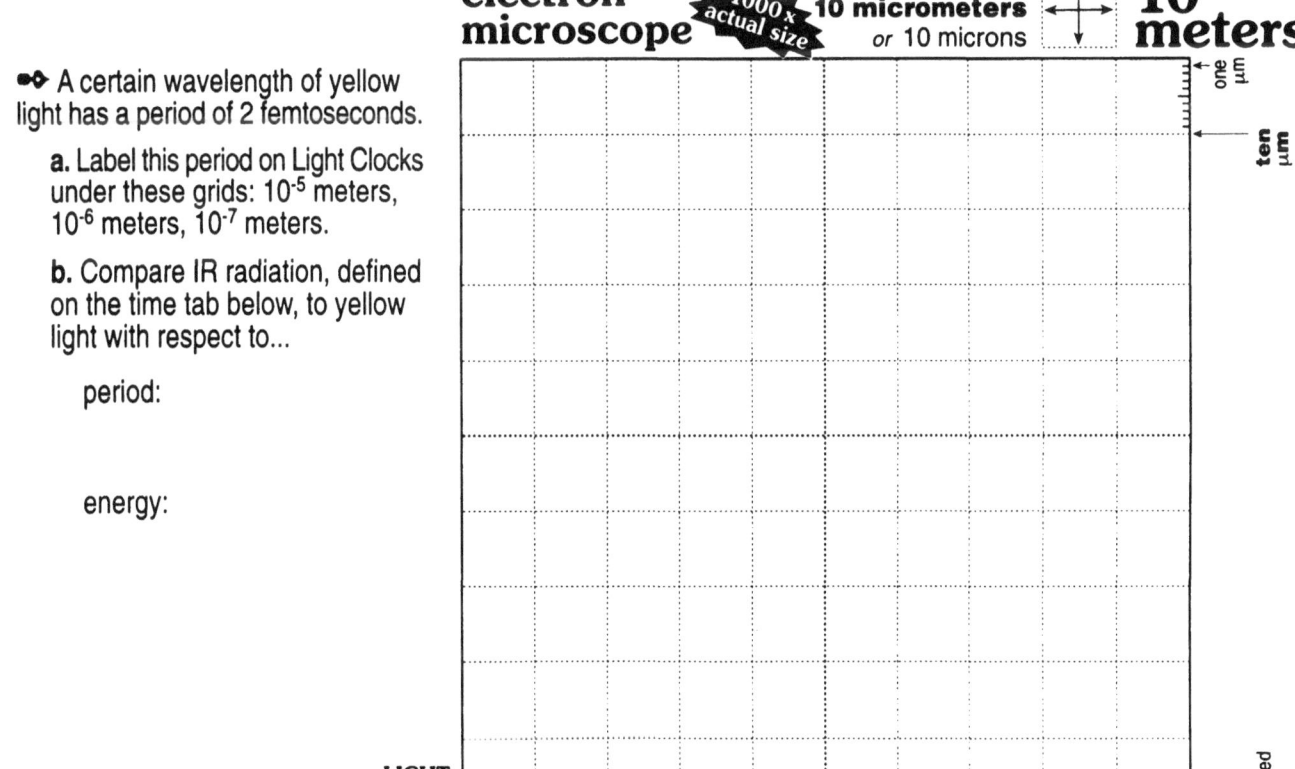

10^{-6} meters

cells

1 cm represents **1 micrometer** or 1 micron

- one hundred nm
- one µm
- ten µm

yellow light

LIGHT CLOCK femtoseconds

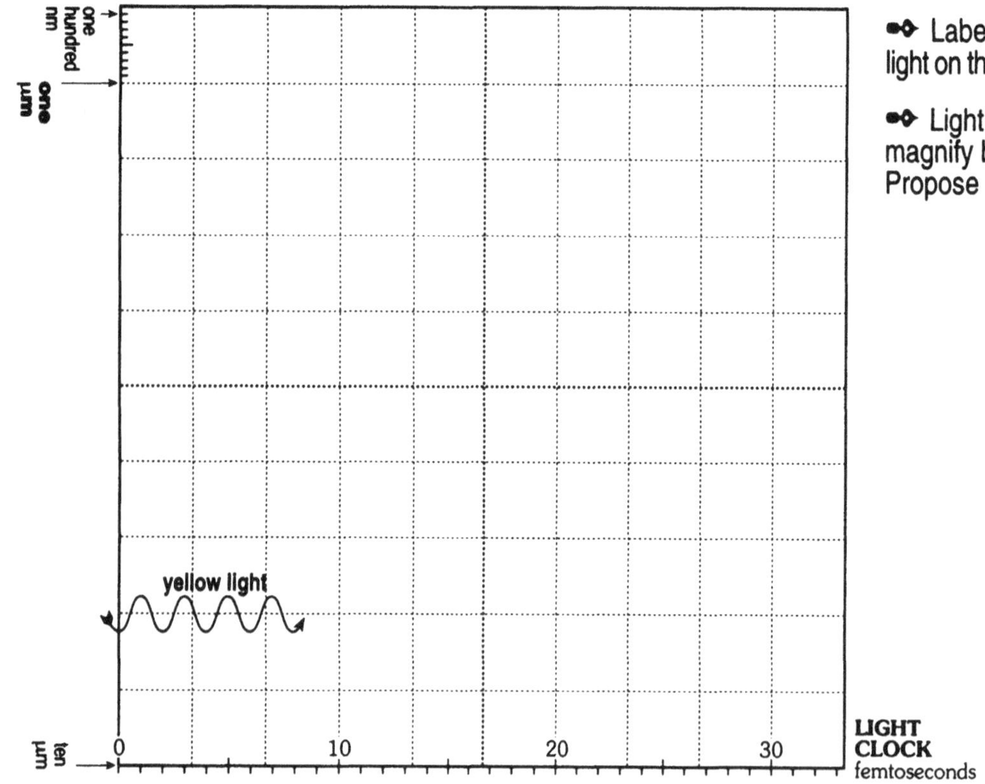

→ Label periods for blue and red light on this Light Clock next to yellow.

→ Light microscopes can generally magnify bacteria, but not viruses. Propose a theory to explain why.

10^{-4} meters

lab microscope

1 cm represents **100 micrometers** or 100 microns

100 x actual size

- ten µm
- one hundred µm
- one mm

1000 fs

LIGHT CLOCK picoseconds

ACTUAL SIZE of area shown in larger window

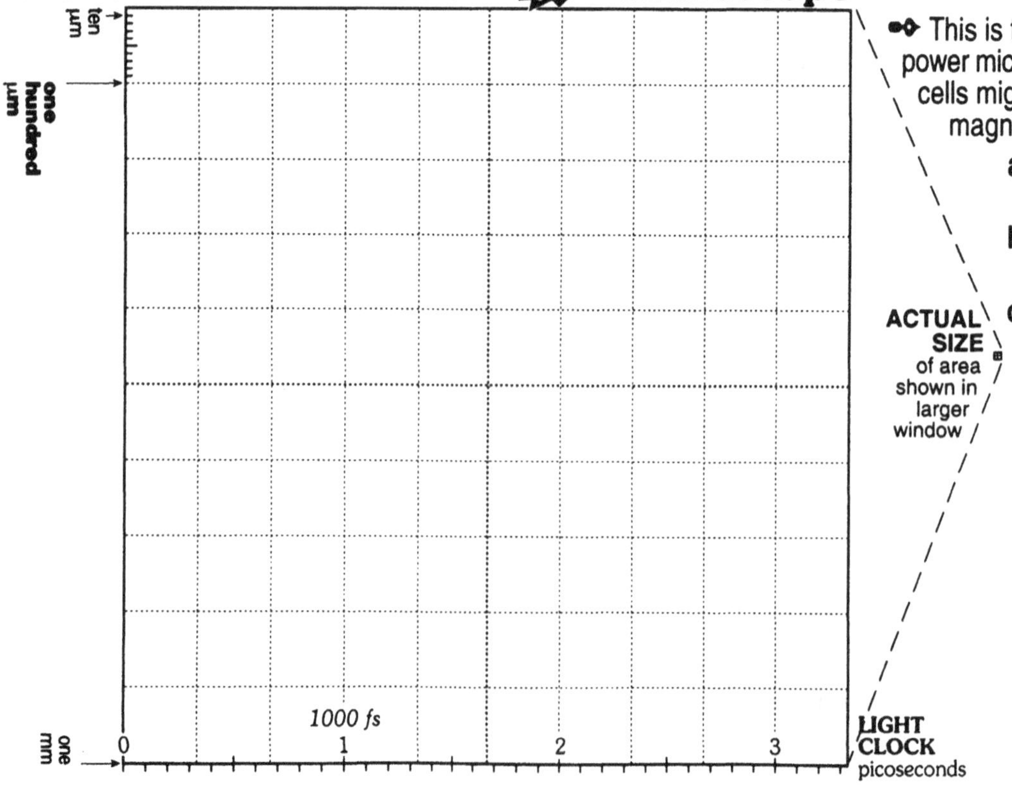

→ This is the visual field of a 100 power microscope. Draw how these cells might look at this magnification:

   a. red blood cells
      diameter = 8.4 microns
   b. human fertilized egg
      diameter = 100 microns
   c. human sperm
      length = 25 microns
      diameter = 1 micron

D: back

→ Compare the wavelength of microwave background radiation to the thickness of paper clip wire.

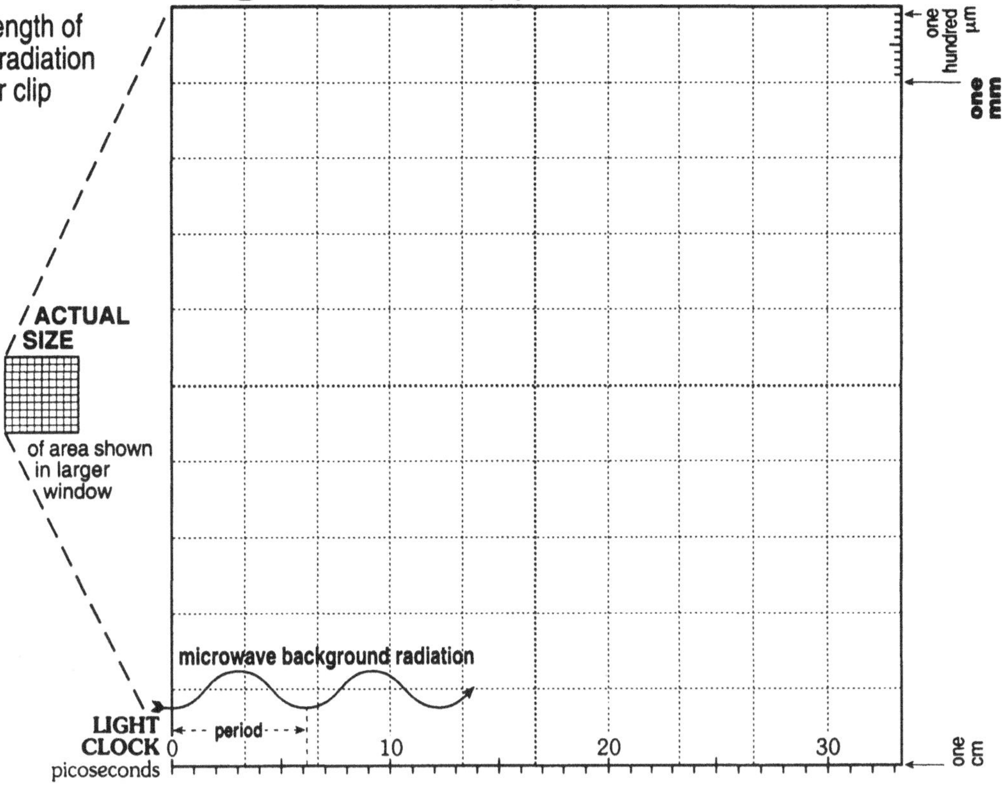

low-power magnifiers — 10× actual size — 1 cm represents **1 millimeter** or 1,000 microns — 10^{-3} **meters**

ACTUAL SIZE of area shown in larger window

microwave background radiation
← period →

LIGHT CLOCK 0 ... 10 ... 20 ... 30
picoseconds

one hundred μm
one mm
one cm

→ Use this Light Clock to estimate how long it takes light to travel each distance:

a. One foot.

b. From thumb to eye with your arm fully extended.

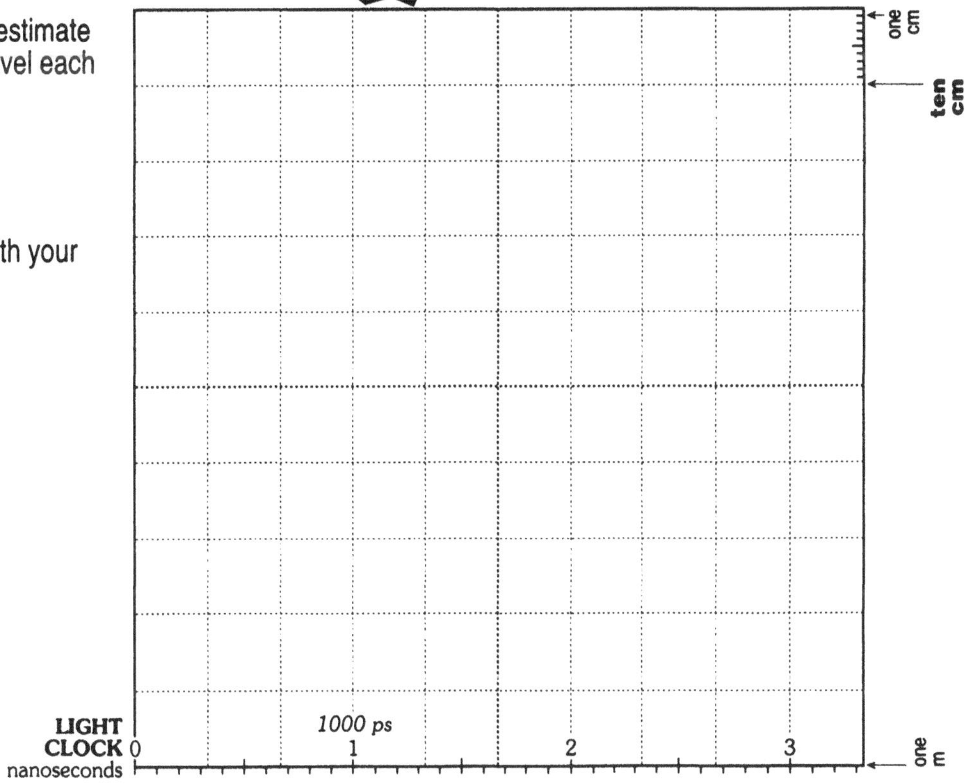

body size — 1/10 actual size — 1 cm represents **10 centimeters** or ≈ 4 inches — 10^{-1} **meters**

LIGHT CLOCK 0 ... 1000 ps ... 1 ... 2 ... 3
nanoseconds

one cm
ten cm
one m

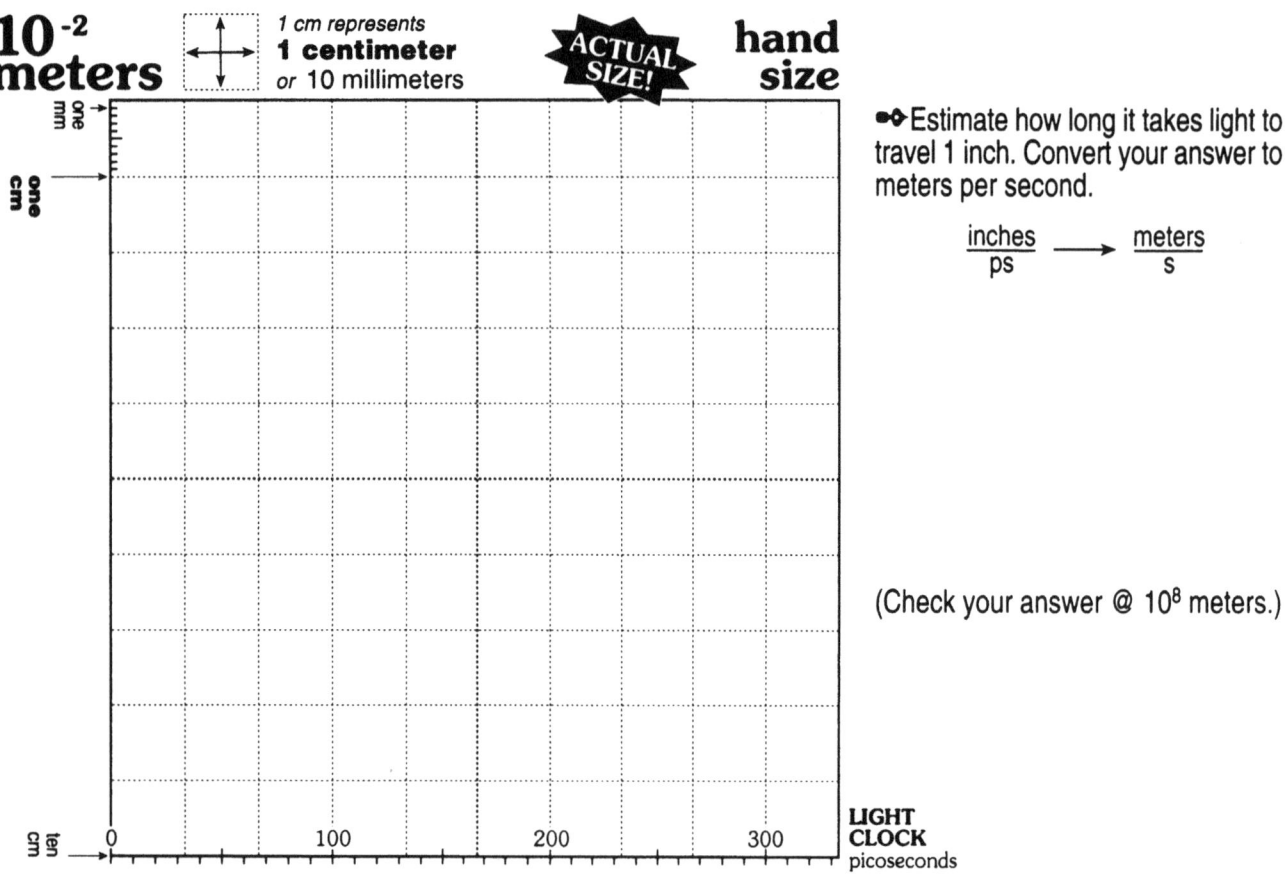

10^{-2} meters

1 cm represents **1 centimeter** or 10 millimeters

ACTUAL SIZE!

hand size

→ Estimate how long it takes light to travel 1 inch. Convert your answer to meters per second.

$$\frac{\text{inches}}{\text{ps}} \longrightarrow \frac{\text{meters}}{\text{s}}$$

(Check your answer @ 10^8 meters.)

LIGHT CLOCK picoseconds

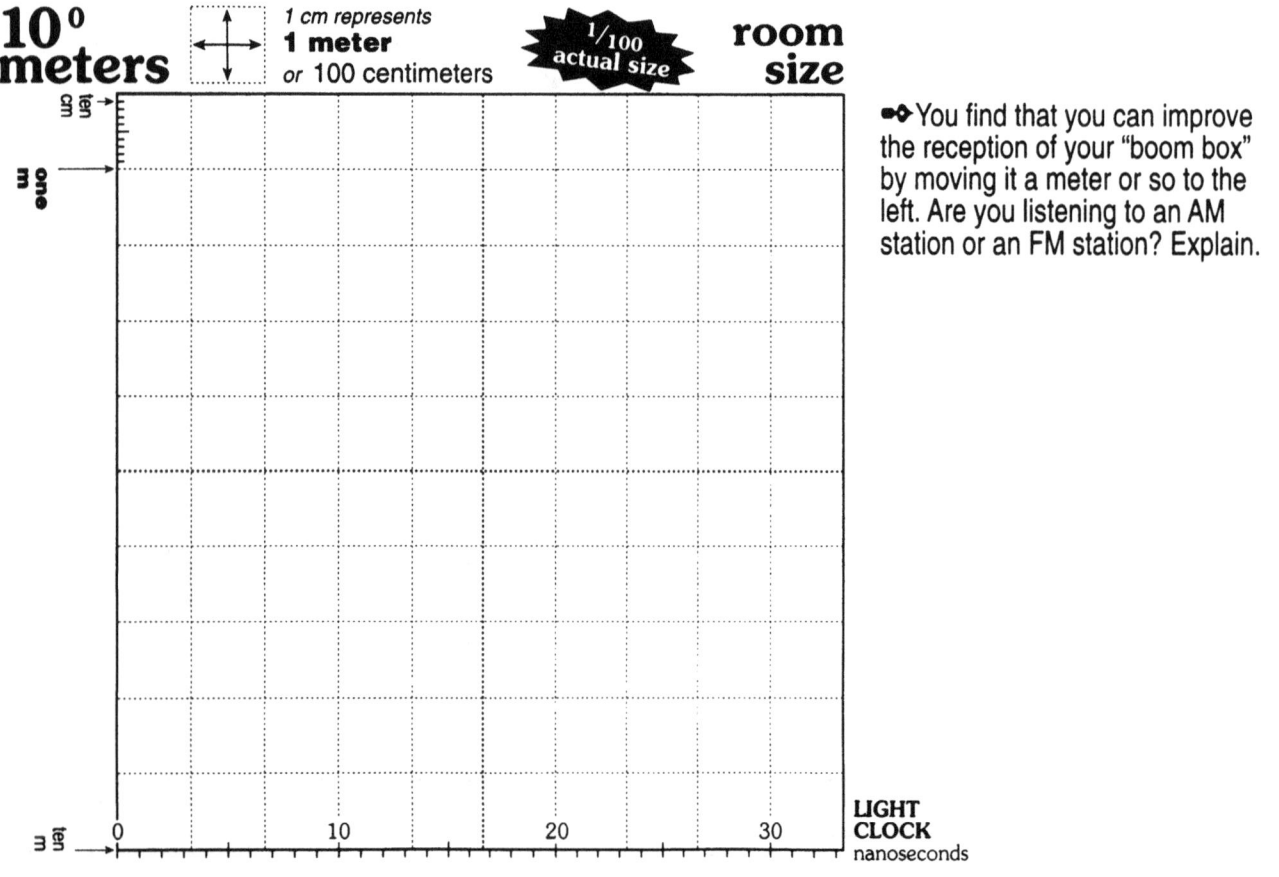

10^0 meters

1 cm represents **1 meter** or 100 centimeters

1/100 actual size

room size

→ You find that you can improve the reception of your "boom box" by moving it a meter or so to the left. Are you listening to an AM station or an FM station? Explain.

LIGHT CLOCK nanoseconds

◆ A radio wave and a track star both run the 100 meter dash.

  a. The track star runs it in 10 seconds. How many nanoseconds is this? (Count pages from 10 seconds to 1 ns.)

  b. How long does it take a radio wave to "dash" 100 meters (cross 10 grid squares)?

  c. How much faster is a radio wave than a track star?

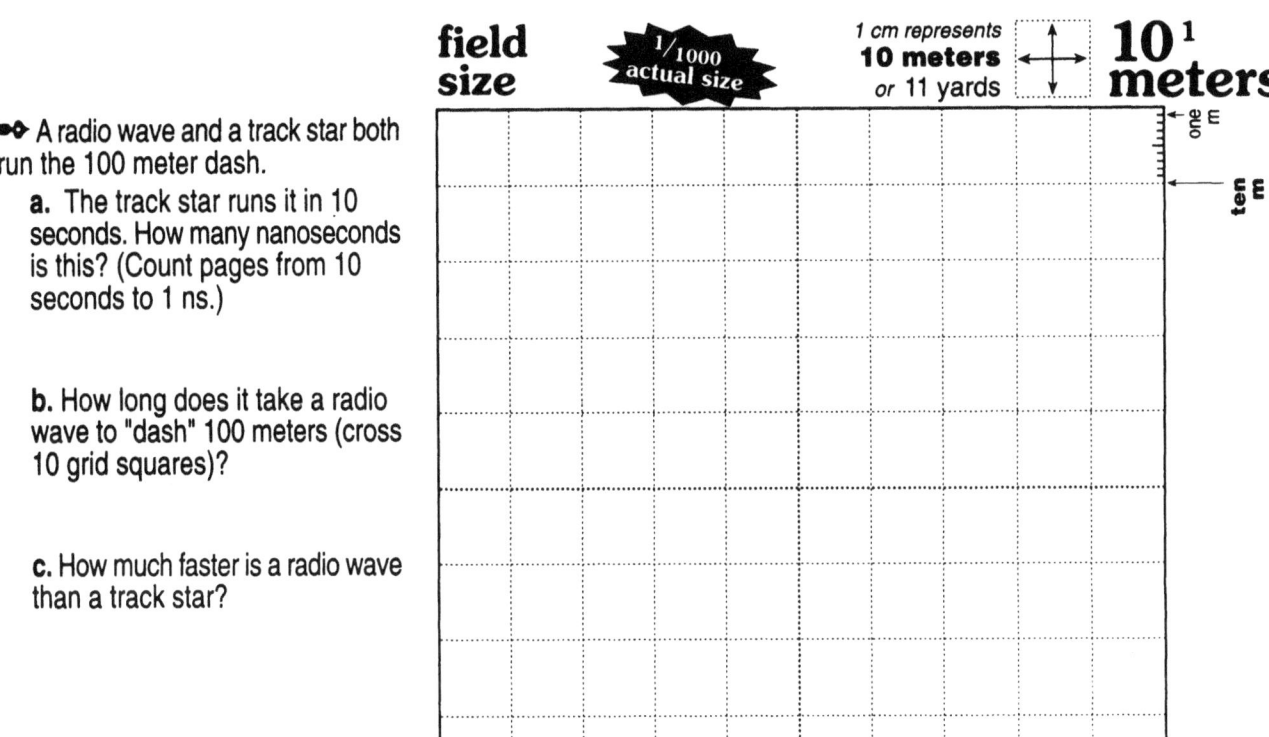

field size $1/1000$ actual size   1 cm represents **10 meters** or 11 yards 10^1 **meters**

LIGHT CLOCK nanoseconds 0   100   200   300

◆ Compare the length of a football field to the height of Mt. Everest:

  a. By rough order of magnitude?

  b. By precise number of football-field lengths that fit the mountain's height?

  Divide numbers:

  Subtract exponents:

  c. By precise order of magnitude? (Find the log of your previous answer to 3 significant figures.)

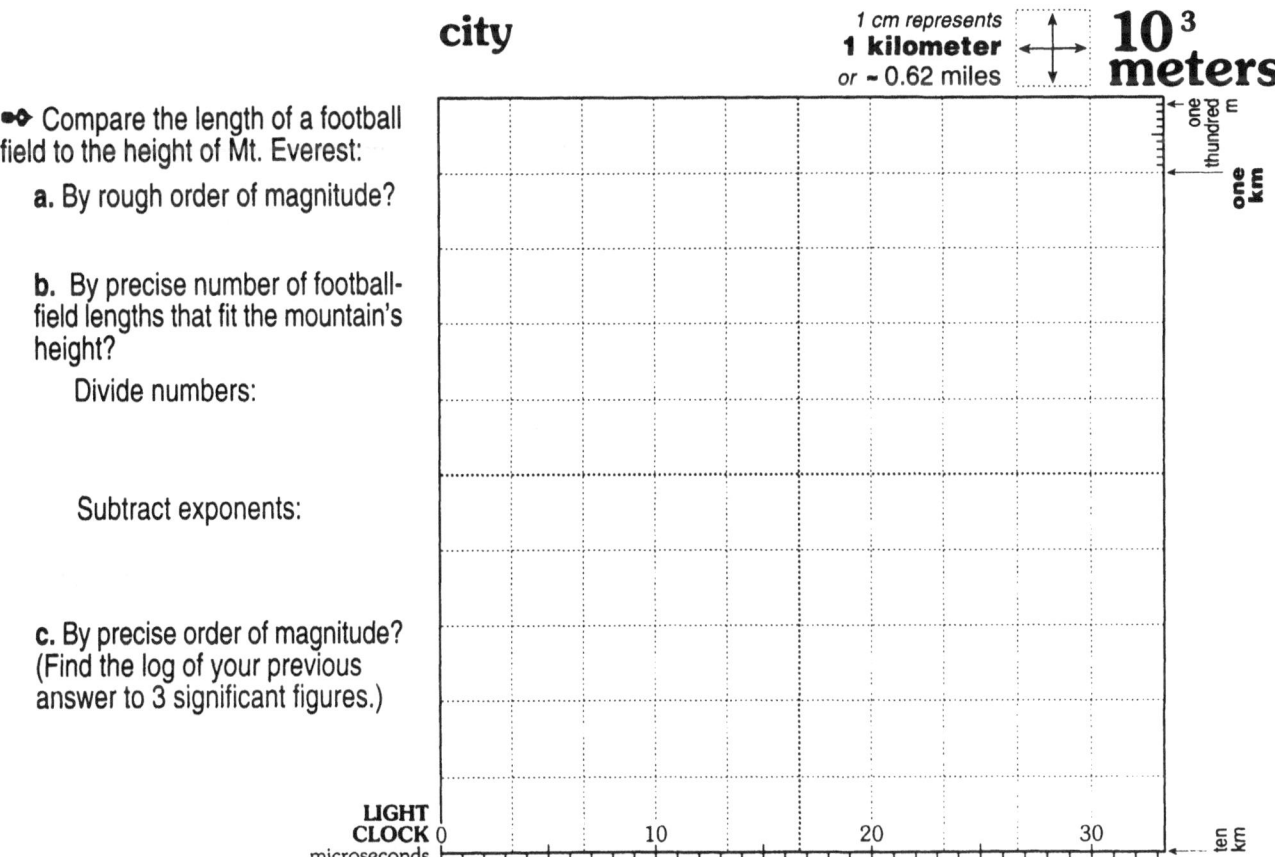

city   1 cm represents **1 kilometer** or ~ 0.62 miles 10^3 **meters**

LIGHT CLOCK microseconds 0   10   20   30

10² meters — neighborhood

1 cm represents **100 meters** or ~ 110 yards

→ You know how far sound travels in 1 second.

a. How long does it take sound to travel a mile? (1 mile = 1.6 km)

b. How can you estimate the distance of lightning (in miles) by the sound of its thunder?

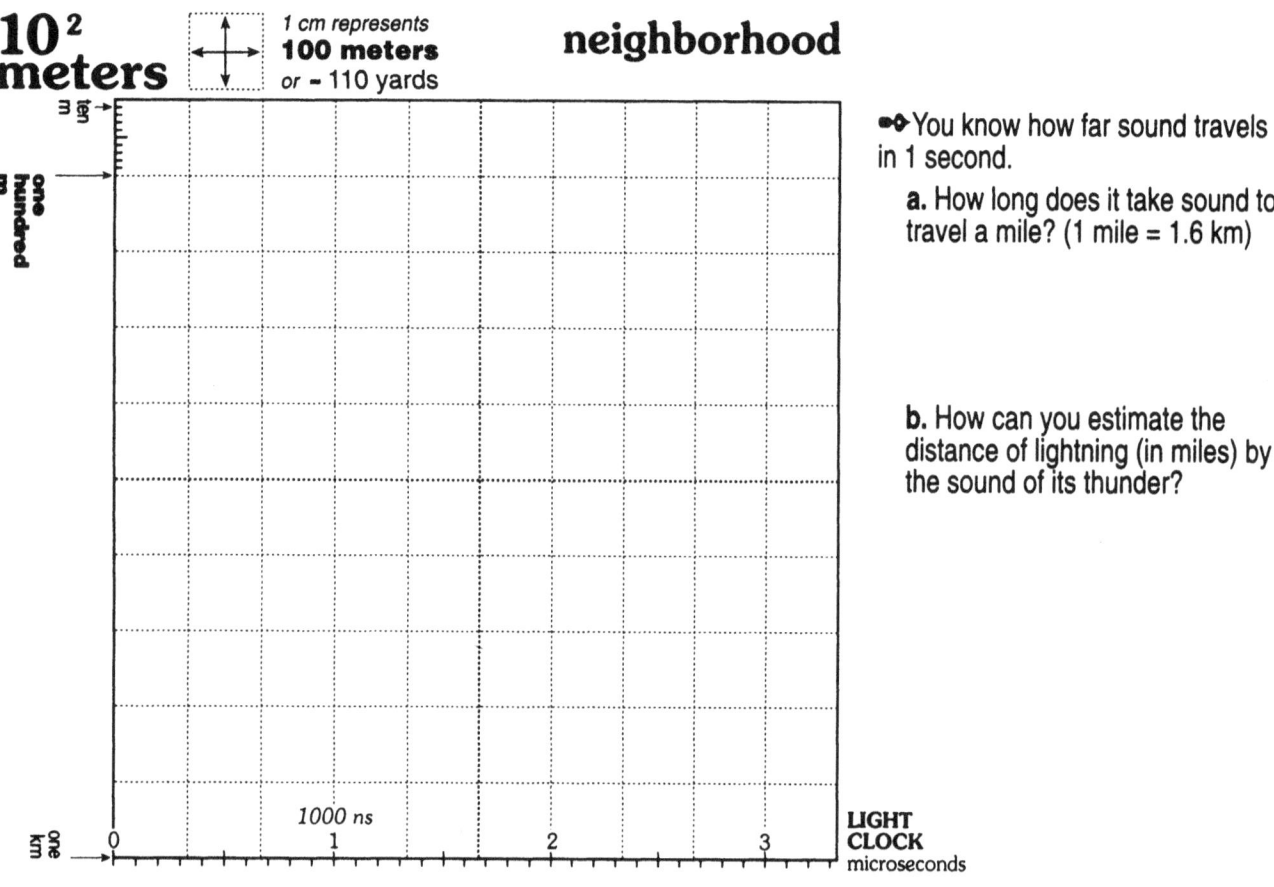

1000 ns

LIGHT CLOCK microseconds

F : back

10⁴ meters — metropolis

1 cm represents **10 kilometers** or ~6.2 miles

→ Radio waves longer than this one are not regulated by the FCC. Anyone can broadcast at longer wavelengths without a license.

a. How long is this wavelength in kilometers? In meters?

b. How many waves whiz by in...

111 µs? _____

333 µs? _____

1 ms? _____

1 s? _____

c. What is the frequency (in cycles per second) of this wave?

d. How far does it move in one second?

e. What is its speed in m/sec? (Check your answer @ 10⁸ meters.)

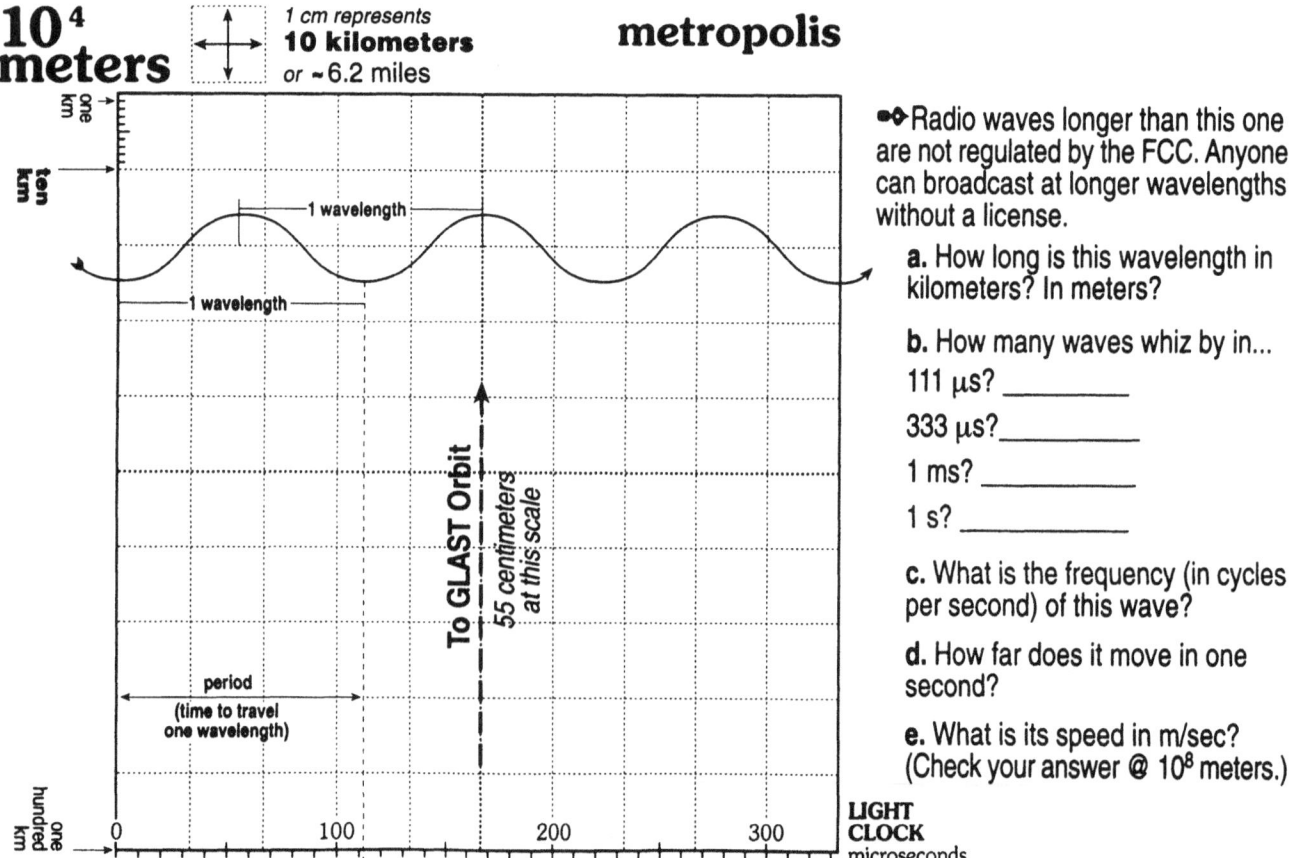

To GLAST Orbit — 55 centimeters at this scale

period (time to travel one wavelength)

LIGHT CLOCK microseconds

◆ If you wanted to draw the moon's orbit at this scale, how tall would you have to make this grid? Explain your reasoning.

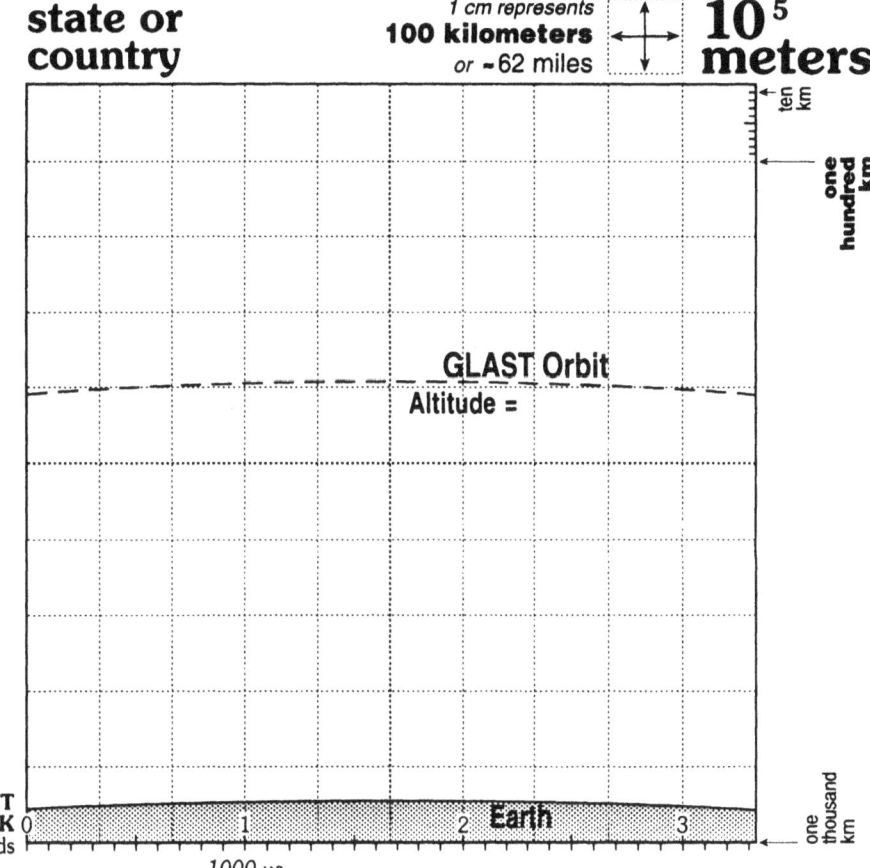

◆ Answer to the nearest power of ten:

a. How many Earths fit across Jupiter?_____

b. How many Jupiters fit across the Sun?_____

c. How many Earths fit across the Sun?_____

Volume varies by the cube of distance. Therefore...

d. How many Earths fit inside Jupiter?_____

e. How many Jupiters fit inside the Sun?_____

f. How many Earths fit inside the Sun?_____

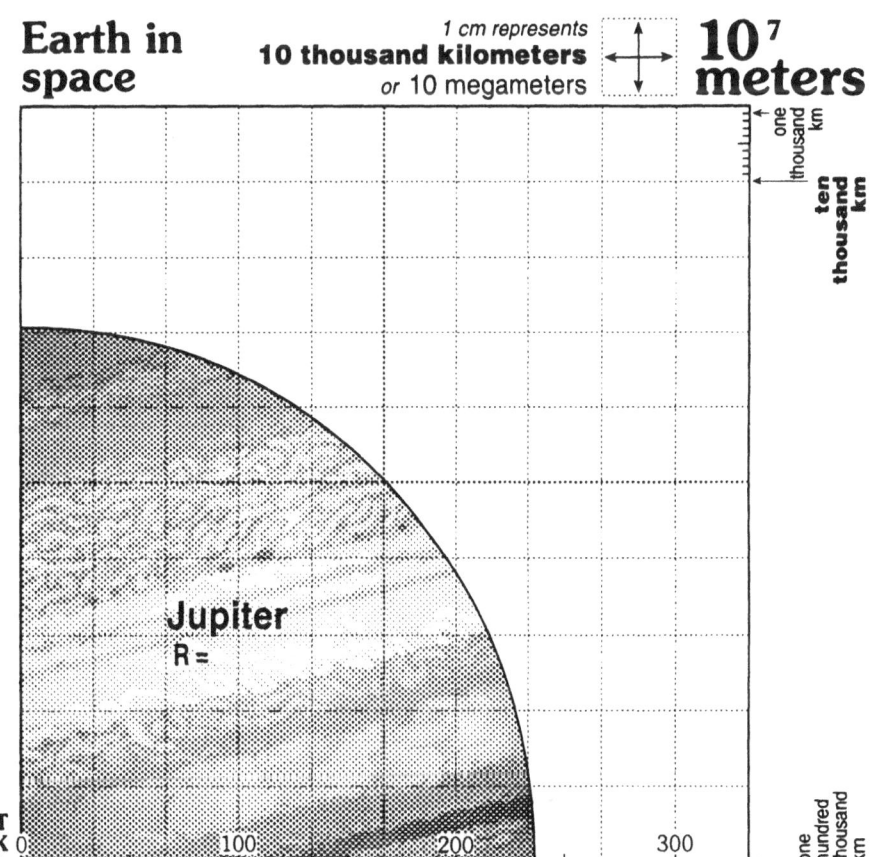

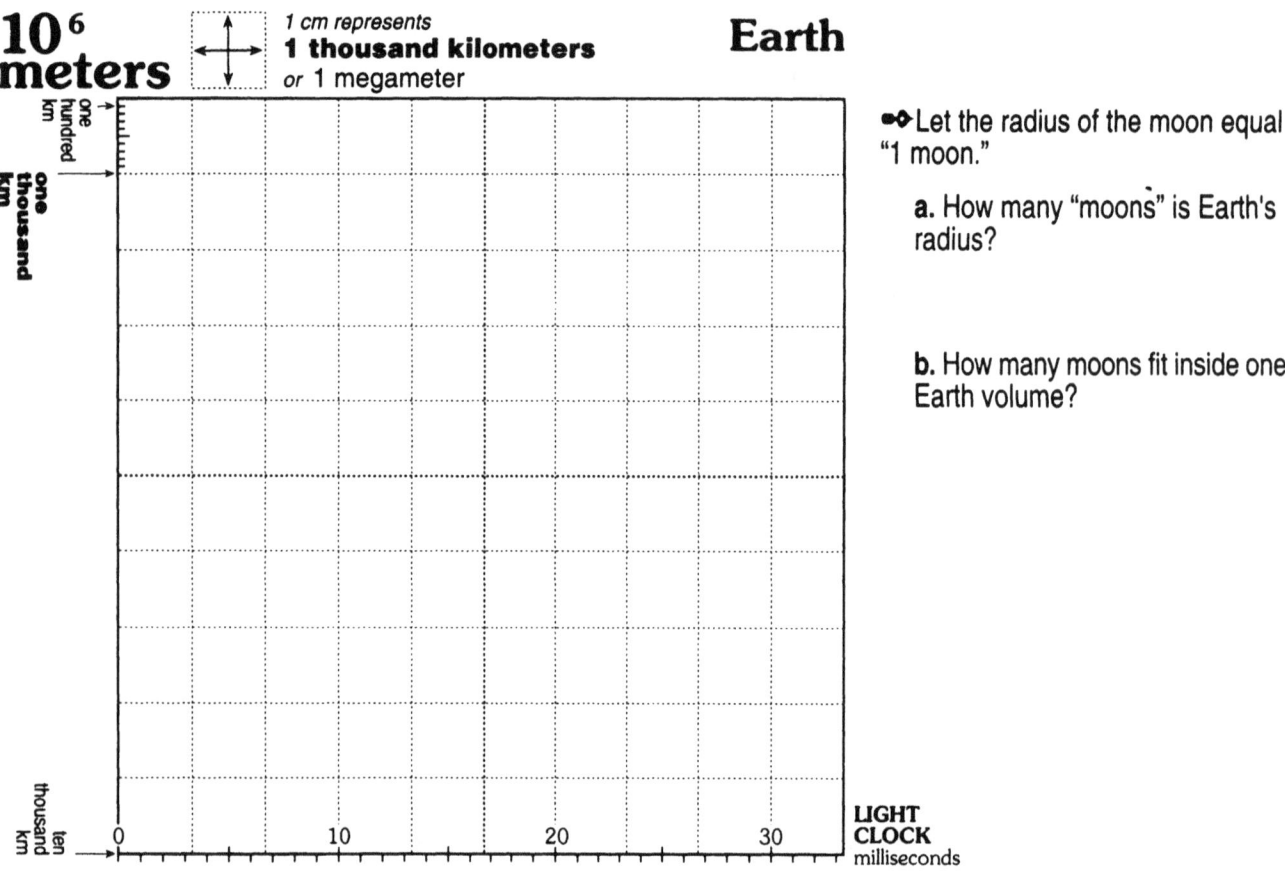

10^6 meters — Earth

1 cm represents **1 thousand kilometers** *or 1 megameter*

LIGHT CLOCK — milliseconds

➤ Let the radius of the moon equal "1 moon."

  a. How many "moons" is Earth's radius?

  b. How many moons fit inside one Earth volume?

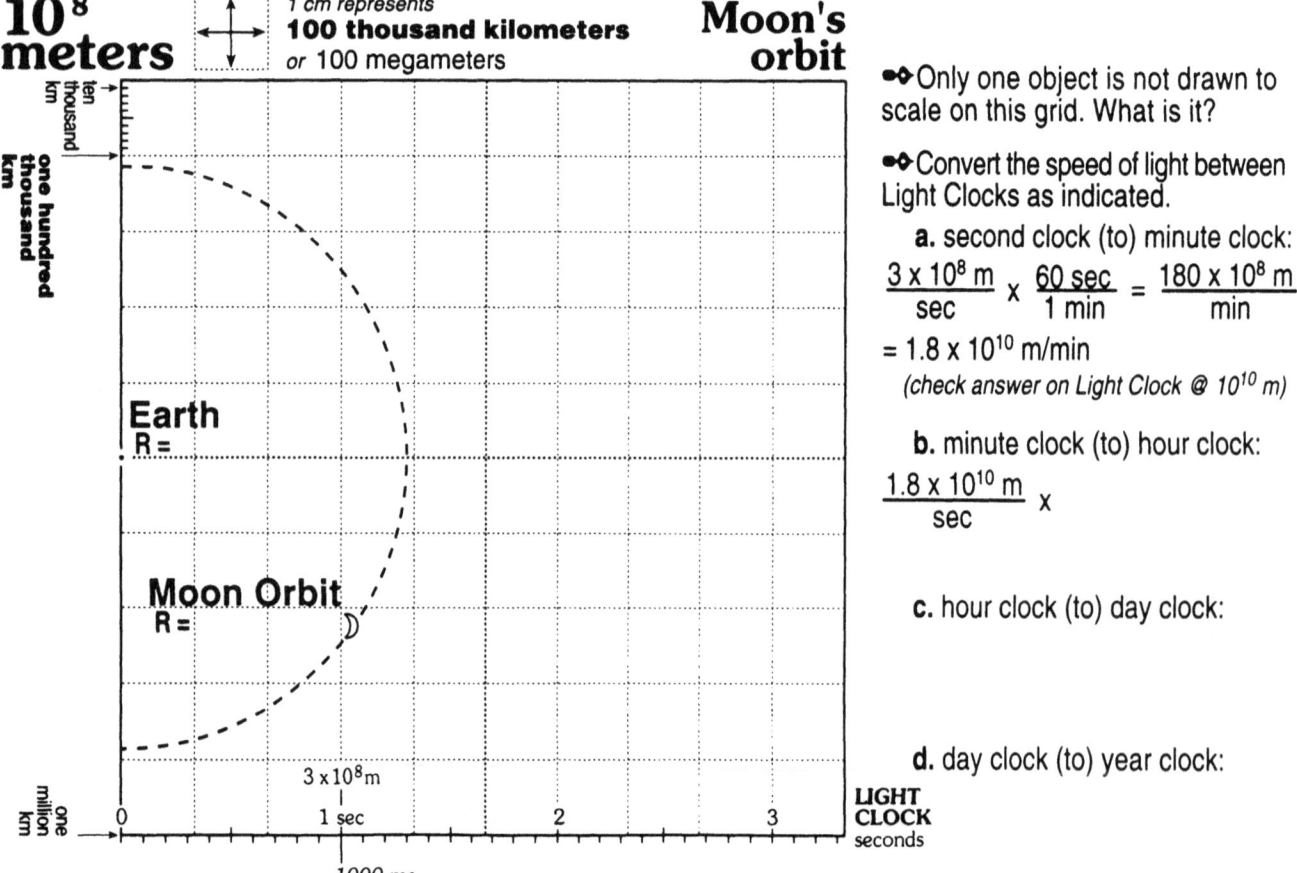

10^8 meters — Moon's orbit

1 cm represents **100 thousand kilometers** *or 100 megameters*

Earth R =

Moon Orbit R =

3×10^8 m / 1 sec

LIGHT CLOCK — seconds

➤ Only one object is not drawn to scale on this grid. What is it?

➤ Convert the speed of light between Light Clocks as indicated.

  a. second clock (to) minute clock:

  $$\frac{3 \times 10^8 \text{ m}}{\text{sec}} \times \frac{60 \text{ sec}}{1 \text{ min}} = \frac{180 \times 10^8 \text{ m}}{\text{min}}$$

  $= 1.8 \times 10^{10}$ m/min

  (check answer on Light Clock @ 10^{10} m)

  b. minute clock (to) hour clock:

  $$\frac{1.8 \times 10^{10} \text{ m}}{\text{sec}} \times$$

  c. hour clock (to) day clock:

  d. day clock (to) year clock:

⚫➔ At about 4 moon orbits from Earth, you reach a place where the gravitational attraction of the Sun and Earth are equal. This is called a **LaGrange Point**.

a. Locate and label this point on the grid.

b. Astronomers observe a small asteroid transiting the sun (crossing its face), 2 million kilometers away. Does it orbit the Earth or the Sun? Explain.

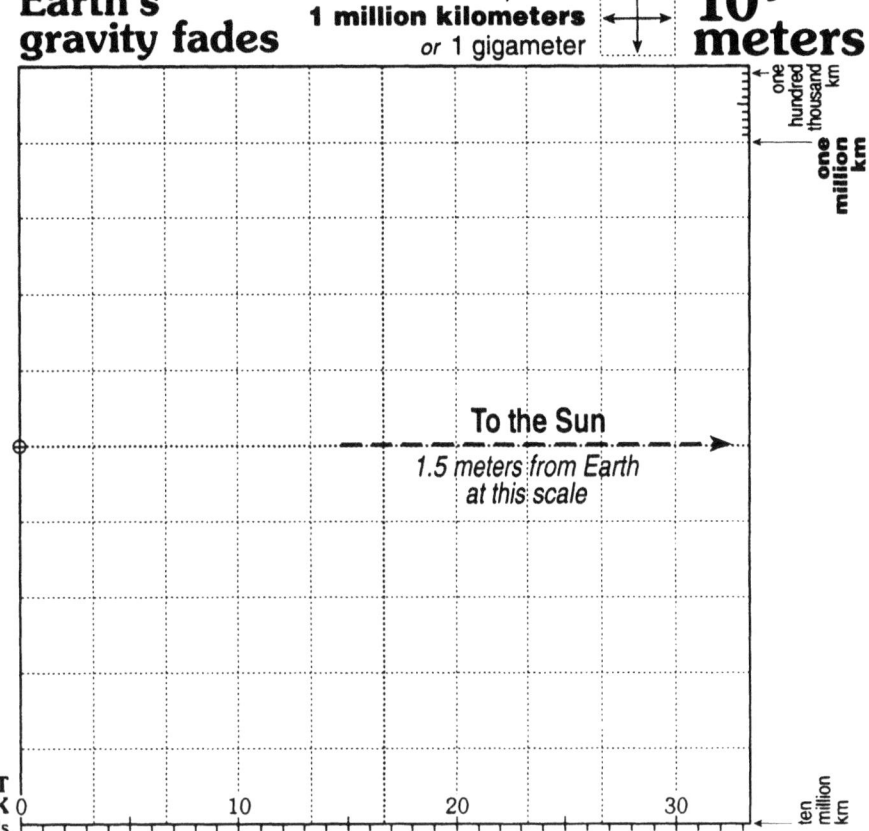

⚫➔ What is an astronomical unit (AU)?

a. How long is 1 AU in kilometers? In meters? In light minutes?

b. How many AUs is Jupiter from the Sun? Show your math.

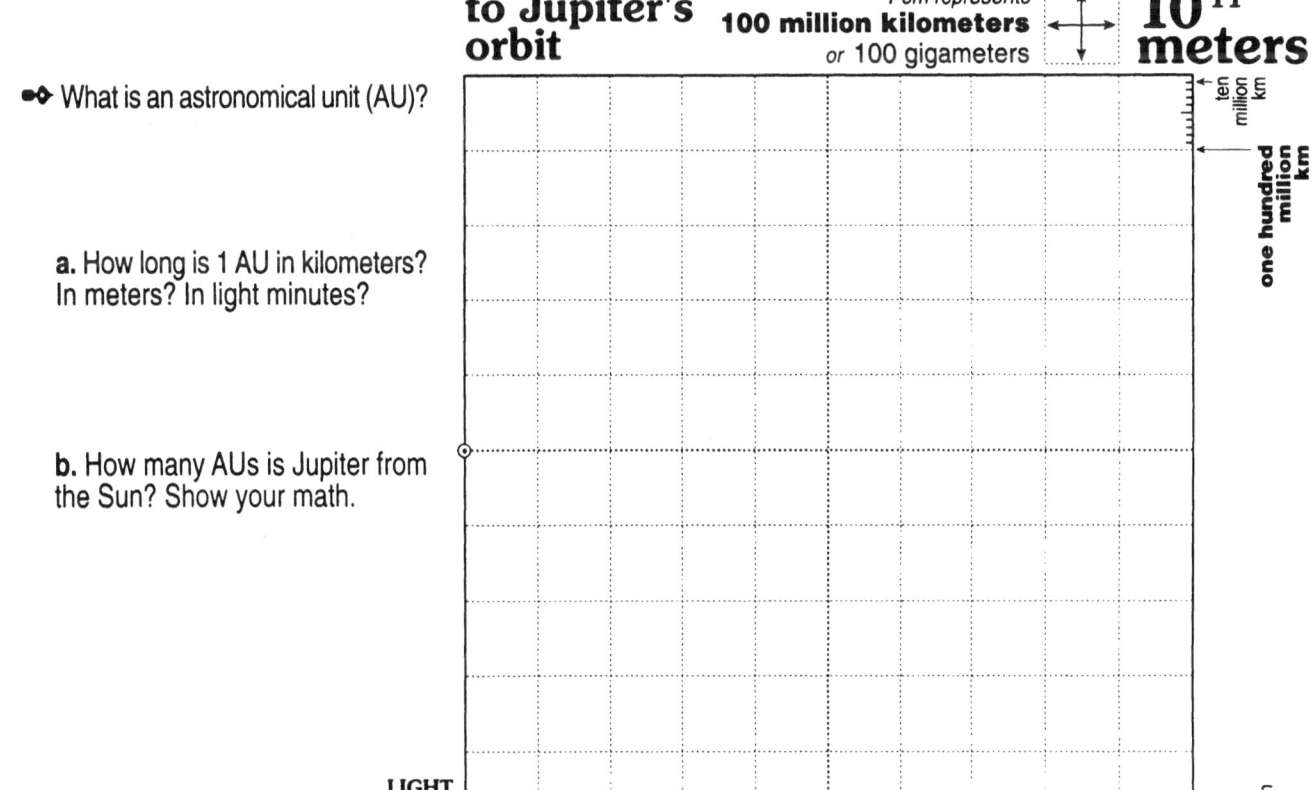

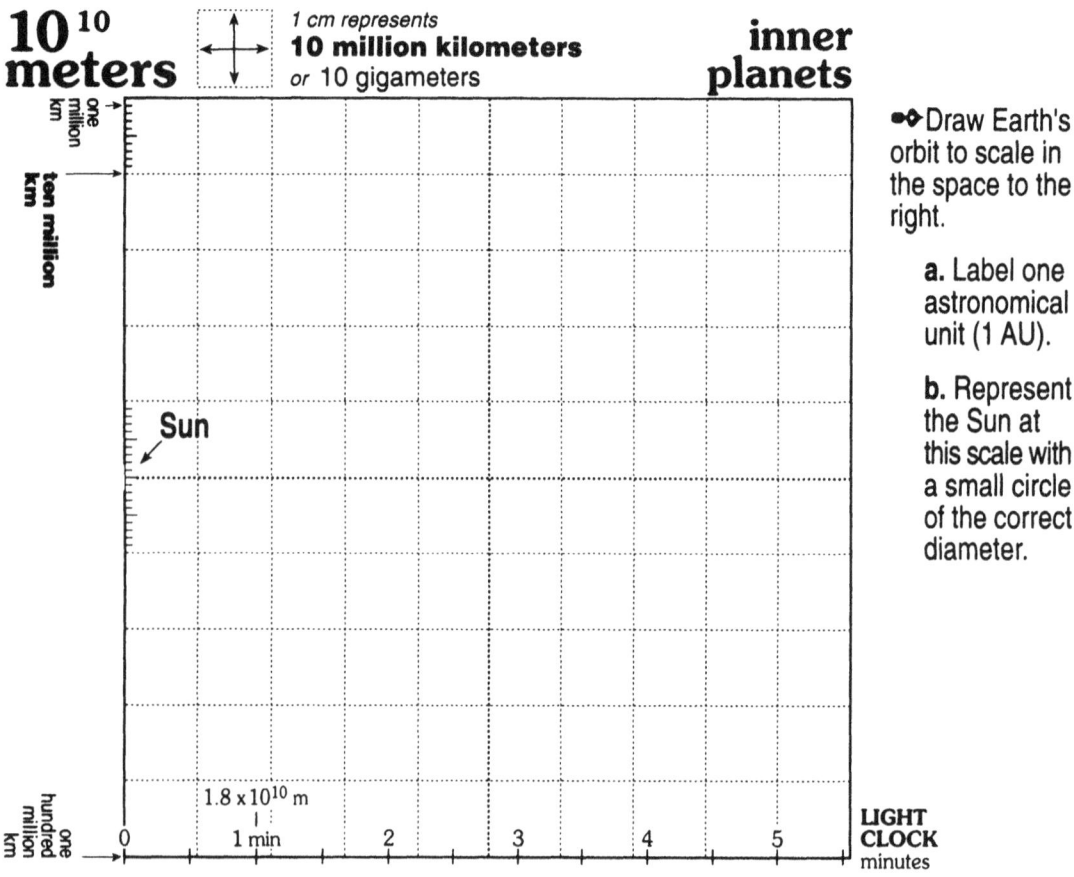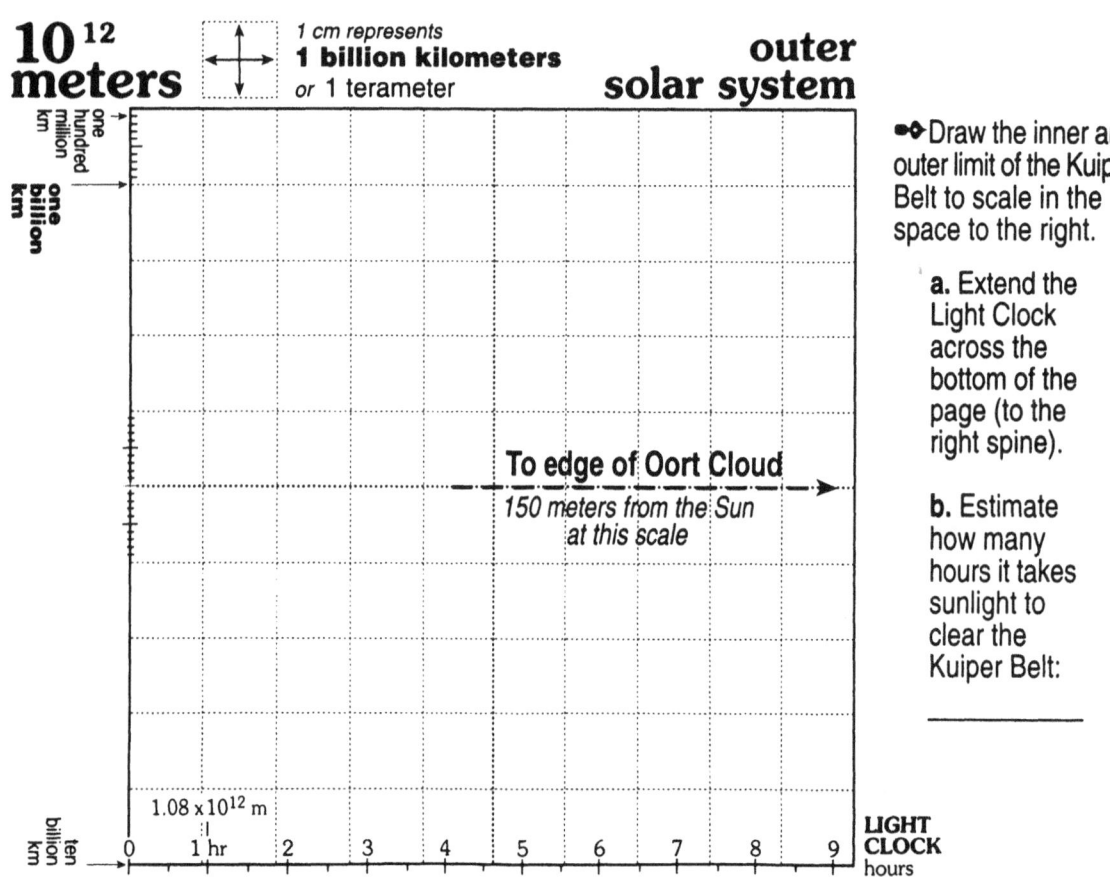

→ The **Kuiper Belt** is a disk-shaped region 4.5 to 15 billion km from the Sun; it is believed to be the source of short-period comets.

   a. Draw this ring to scale on the grid and shade it in.

   b. Estimate how long it takes a gamma ray emitted during a solar flare to clear the outer edge of this ring.

   c. Draw the outer limit of this ring to scale @ 10^{14} meters.)

Please return to questions @ 10^{15} m.

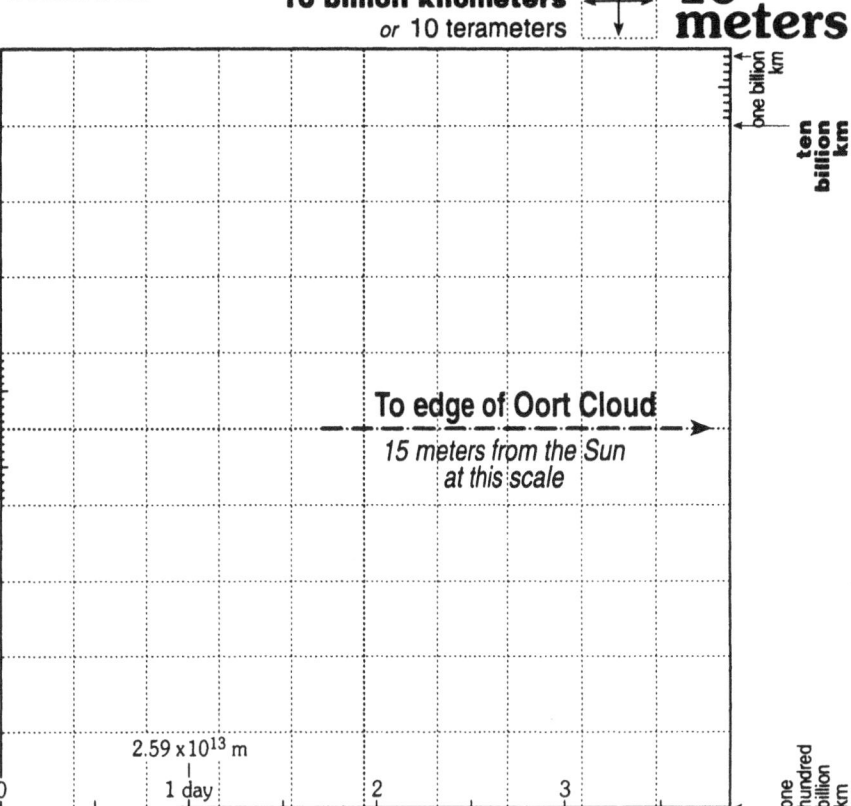

Please complete questions @ 10^{13} m first. Then return here:

→ At this scale, the Kuiper Belt is a barely visible point. Inside this point is our Sun, Earth and entire solar system!

   a. What is the radius of this point in centimeters?

   b. What is the radius of the Oort Cloud at this scale in centimeters?

   c. Compare the outer limits of the Kuiper Belt and Oort Cloud.

→ The periods of comet Hale-Bopp and comet Halley are 4,000 years and 80 years, respectively. Speculate about their orbits around the Sun.

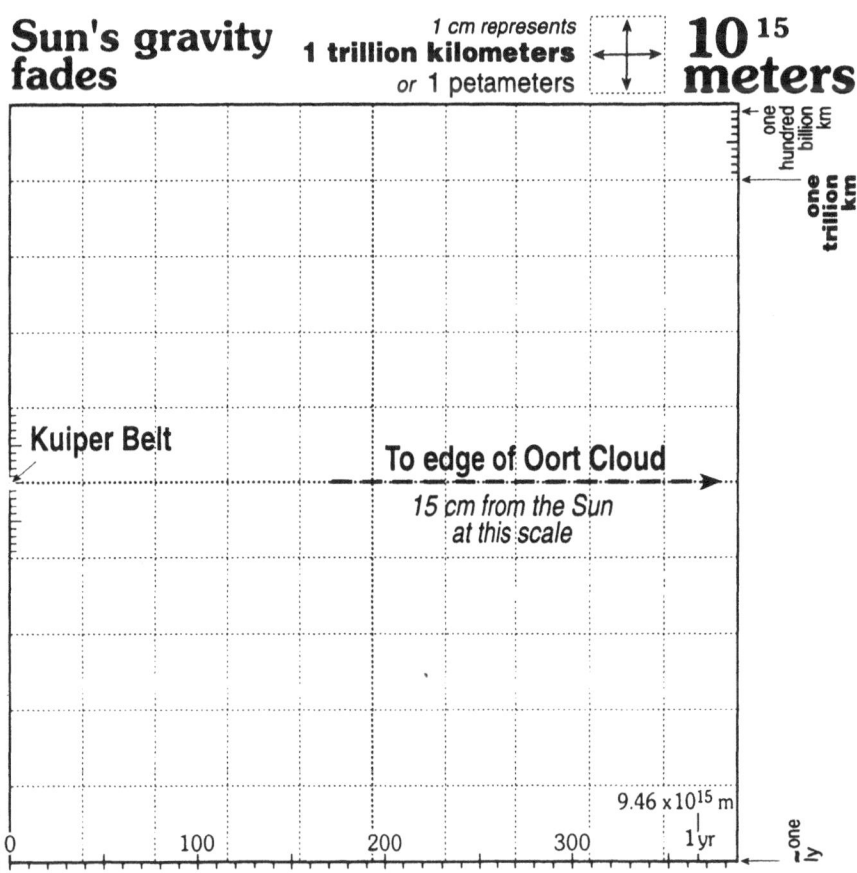

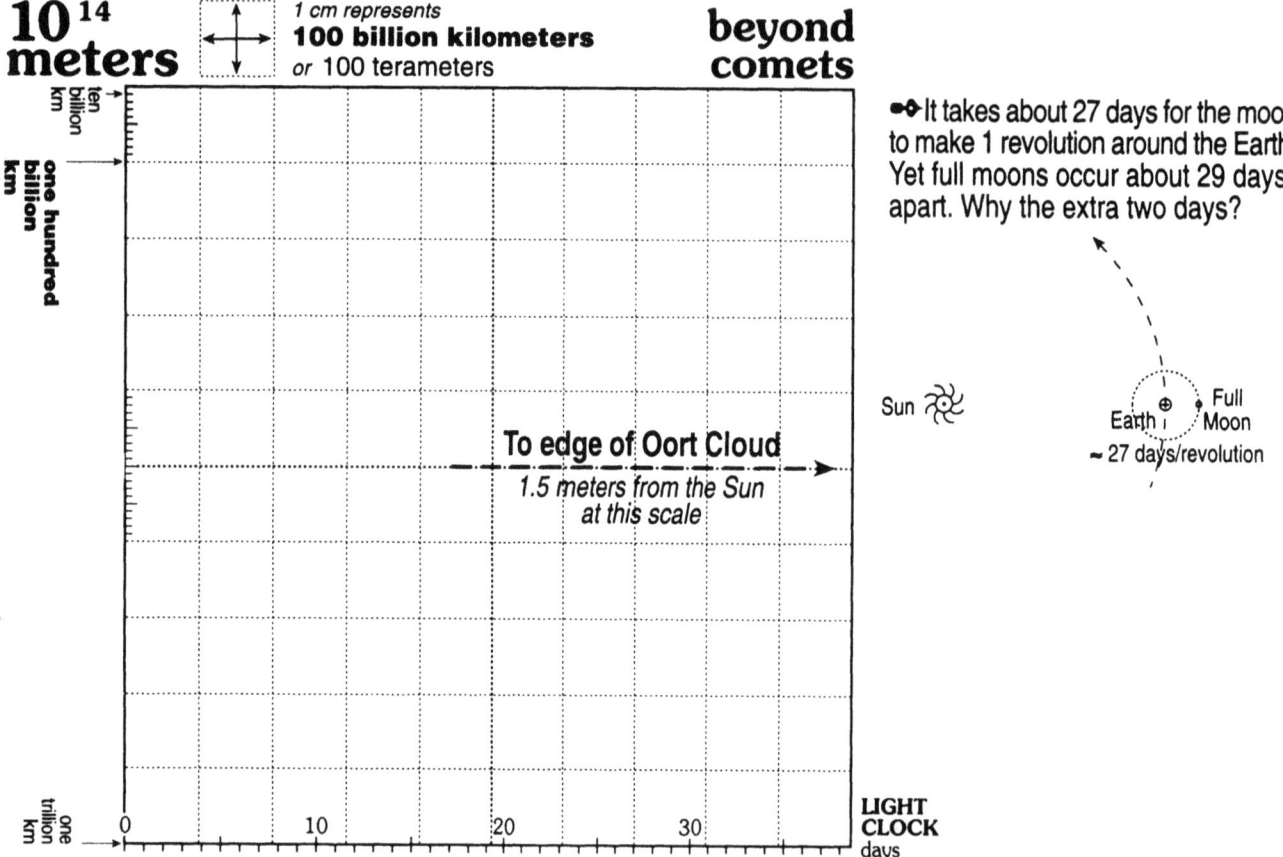

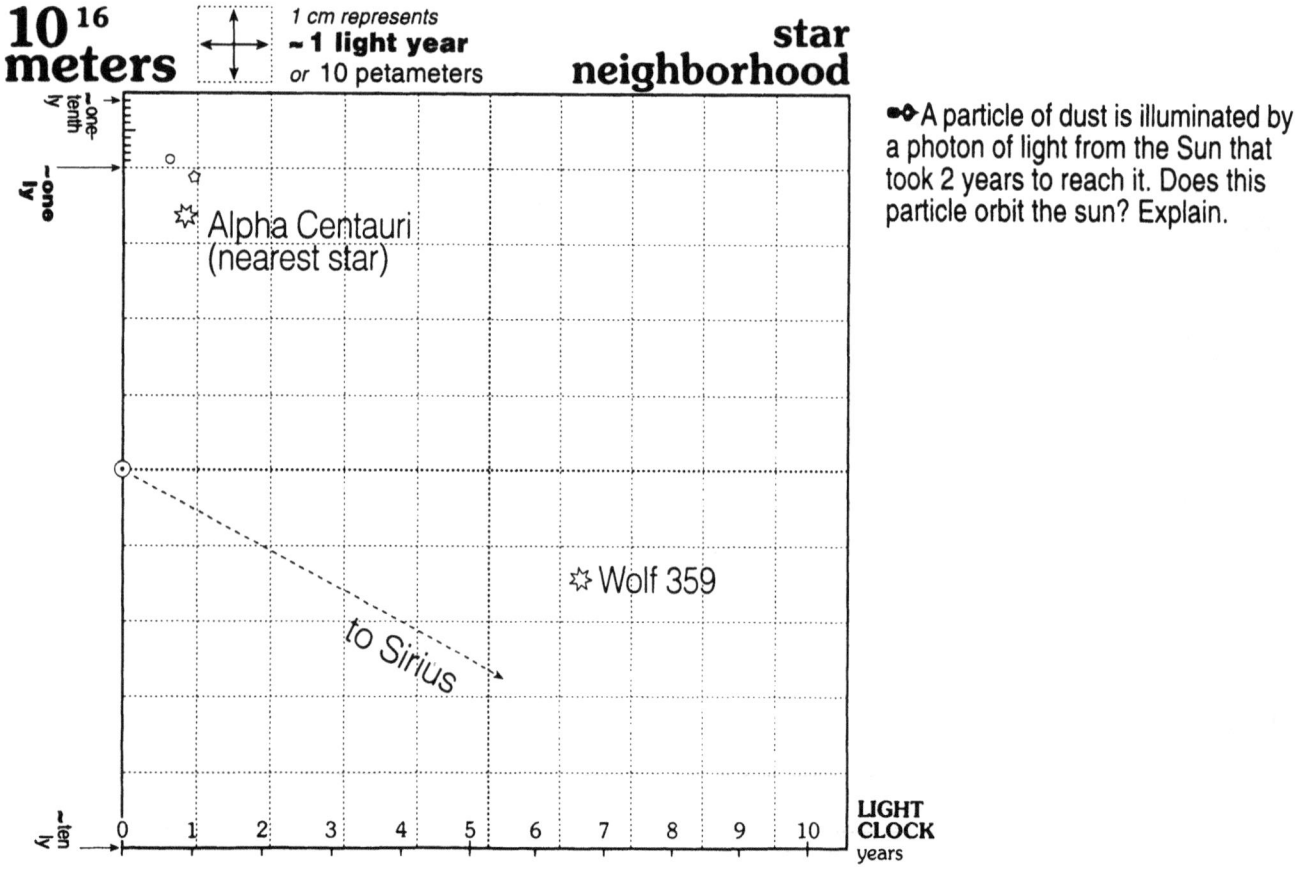

◆◇ Guglielmo Marconi invented radio broadcasting in the 1890's.

a. When were radio signals broadcast on Earth that are currently reaching planets that orbit HD 70642?

b. No patterned radio signals have been discovered so far from HD 70642. What are the implications?

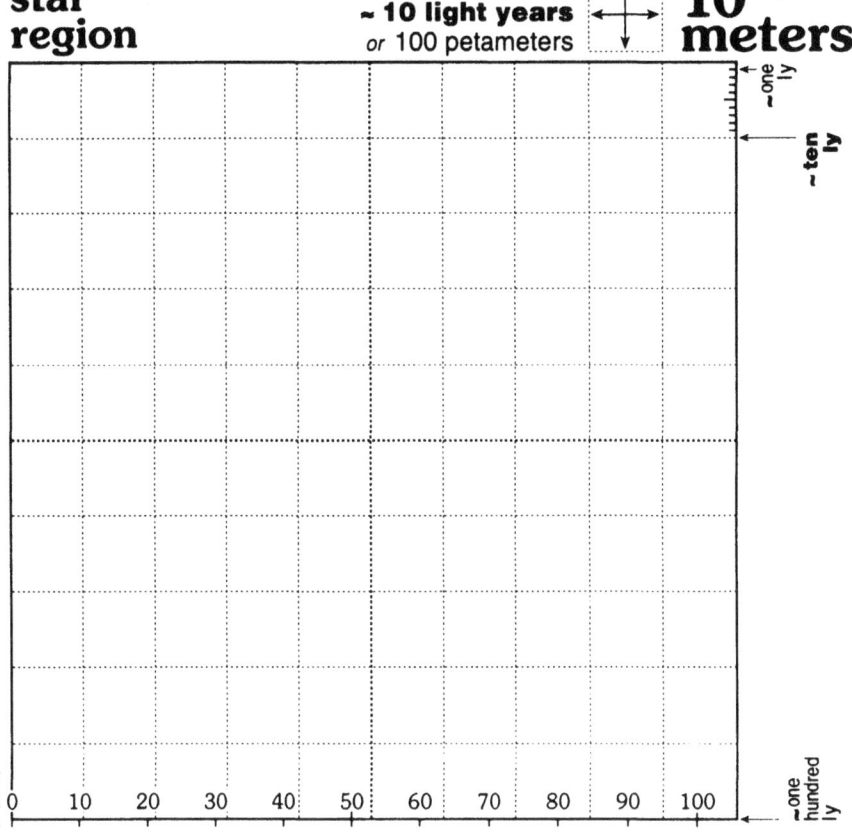

star region — 1 cm represents ~ 10 light years or 100 petameters — 10^{17} meters

LIGHT CLOCK years: 0, 10, 20, 30, 40, 50, 60, 70, 80, 90, 100

~one ly, ~ten ly, ~one hundred ly

◆◇ An astronomer proposes to track gamma-ray emissions from the Crab Pulsar over a period of one continuous hour. Why wouldn't GLAST administrators be receptive to this proposal? (See Time Tab @ 10^{11} meters.)

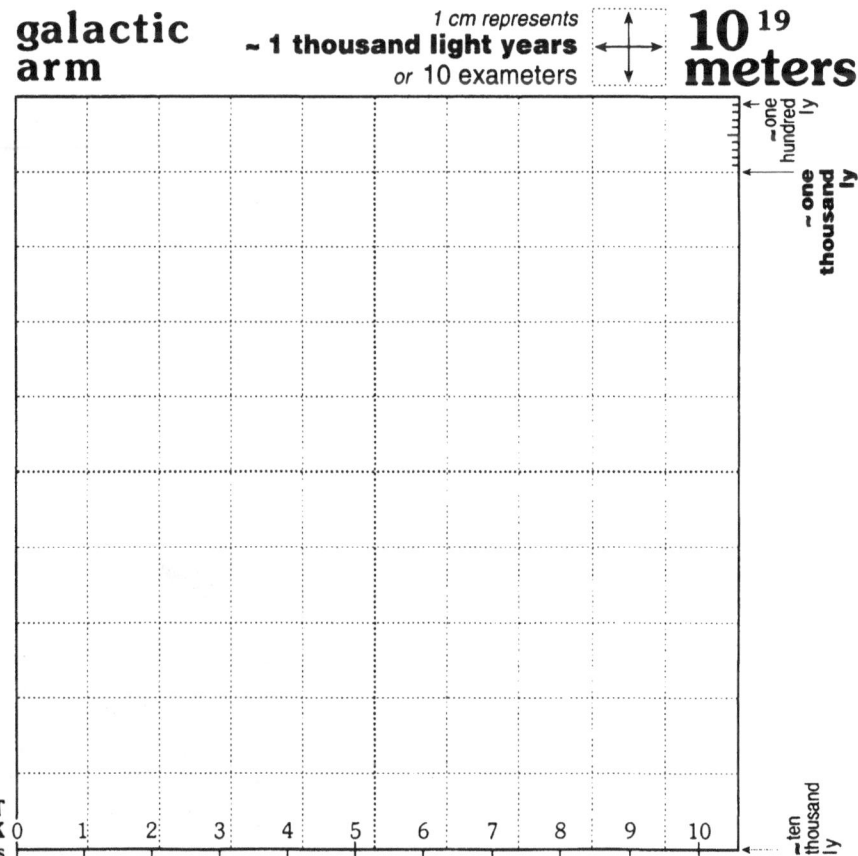

galactic arm — 1 cm represents ~ 1 thousand light years or 10 exameters — 10^{19} meters

LIGHT CLOCK thousand years: 0, 1, 2, 3, 4, 5, 6, 7, 8, 9, 10

~one hundred ly, ~one thousand ly, ~ten thousand ly

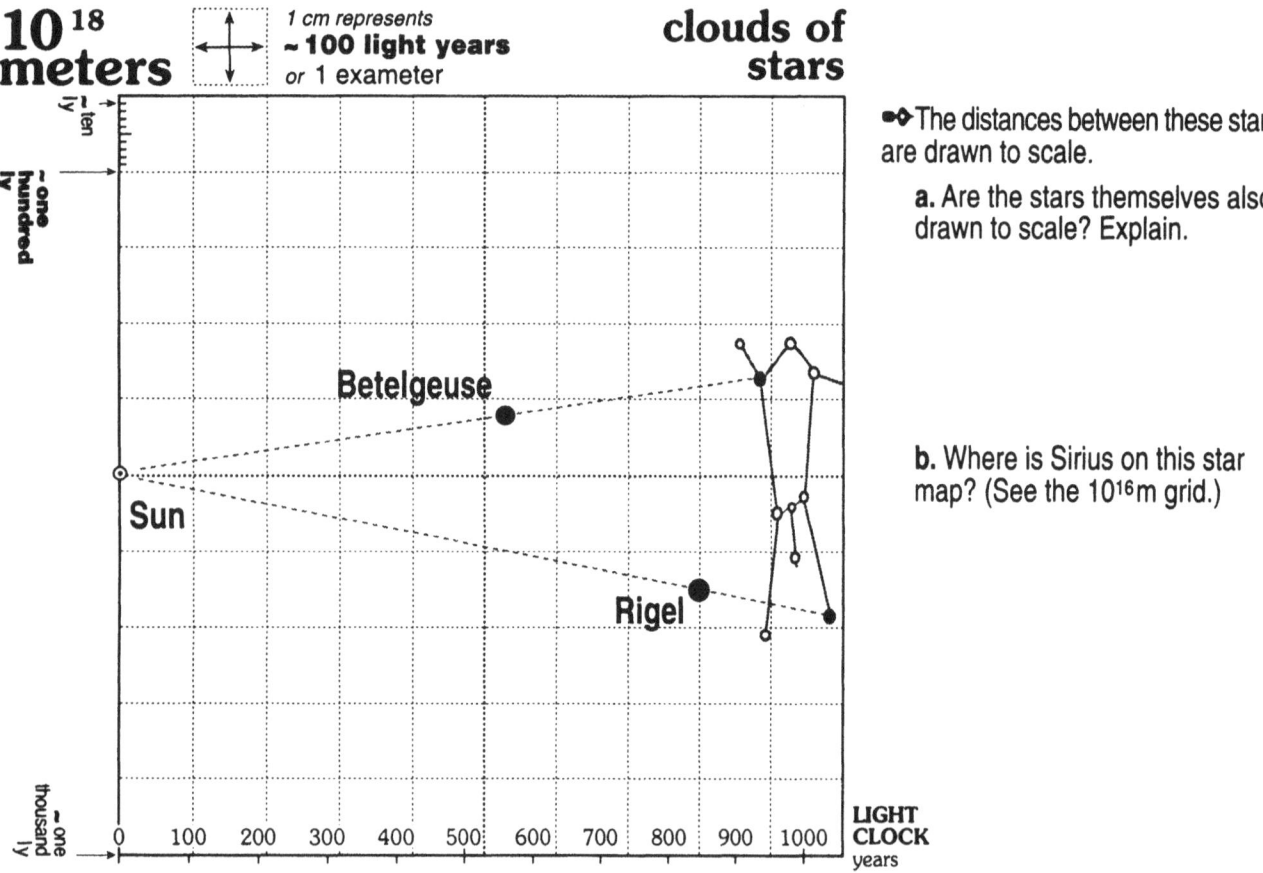

→ The distances between these stars are drawn to scale.

a. Are the stars themselves also drawn to scale? Explain.

b. Where is Sirius on this star map? (See the 10^{16}m grid.)

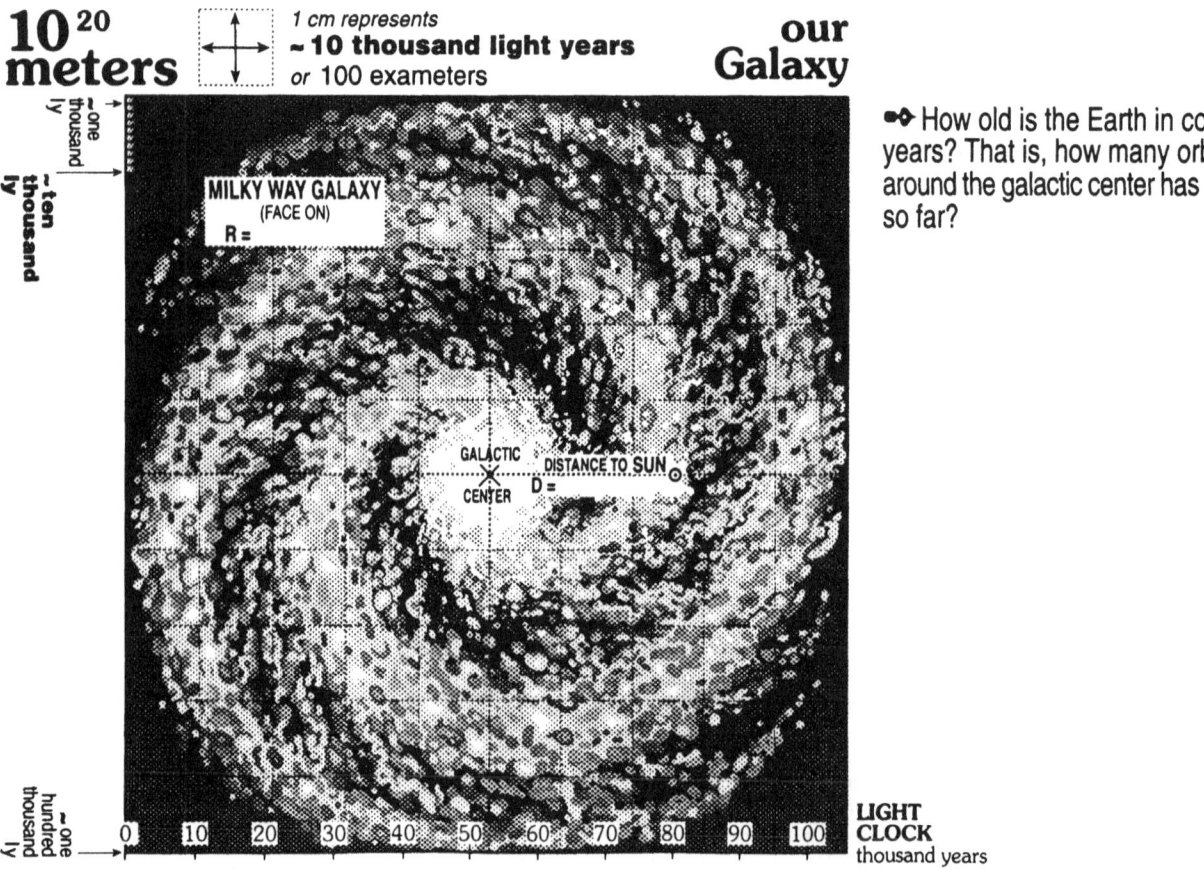

→ How old is the Earth in cosmic years? That is, how many orbits around the galactic center has it made so far?

↦ Andromeda, our galactic neighbor, is about as big as a nickel at this scale. Where is this nickel relative to the Milky Way and Large Magellanic Cloud drawn here?

    a. Describe its location.

    b. Represent its location on this grid with a labeled arrow.

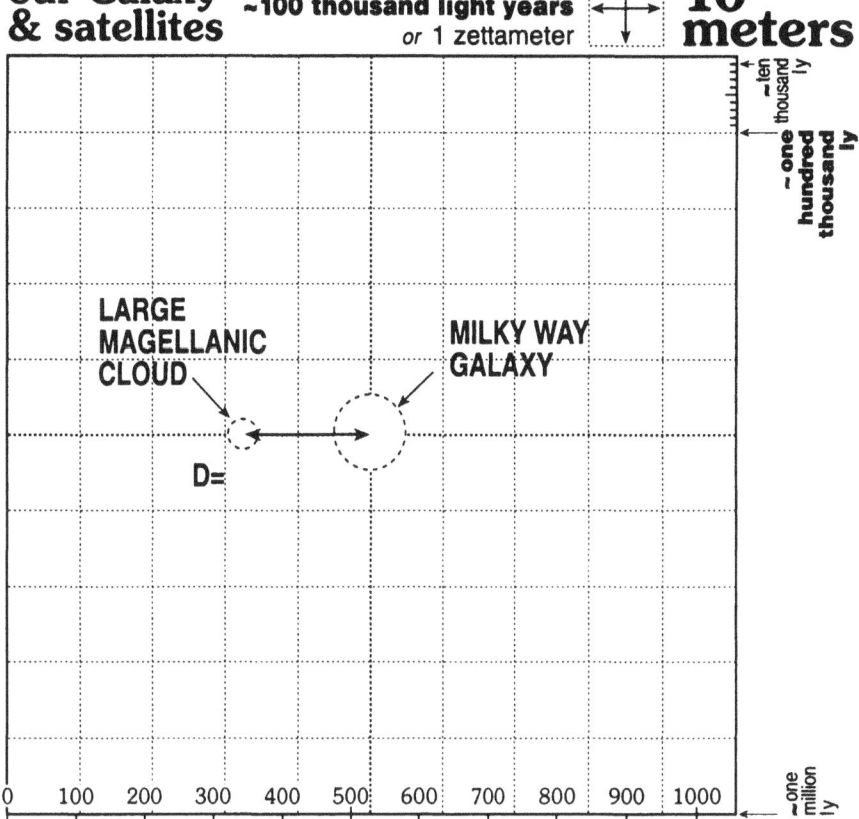

↦ Suppose we someday detect non-random radio waves originating inside the Virgo Group, centered in the Virgo Super Cluster. Could a Virgo civilization also pick up our broadcast waves? Explain.

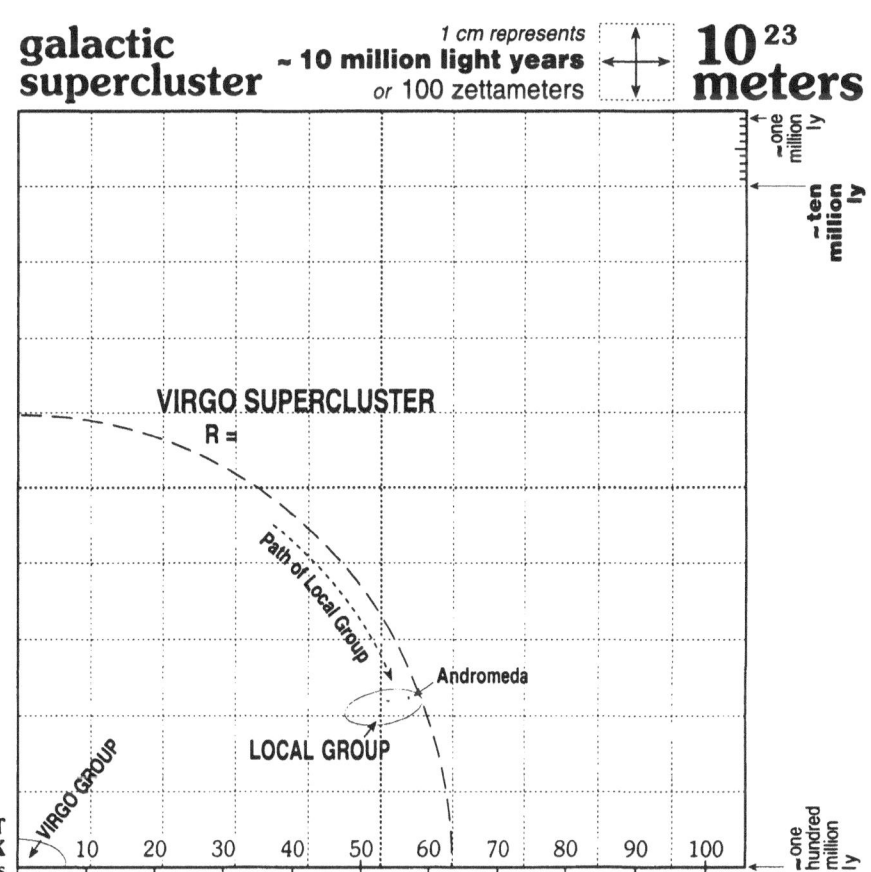

10^{22} meters — local galactic group

1 cm represents ~1 million light years or 10 zettameters

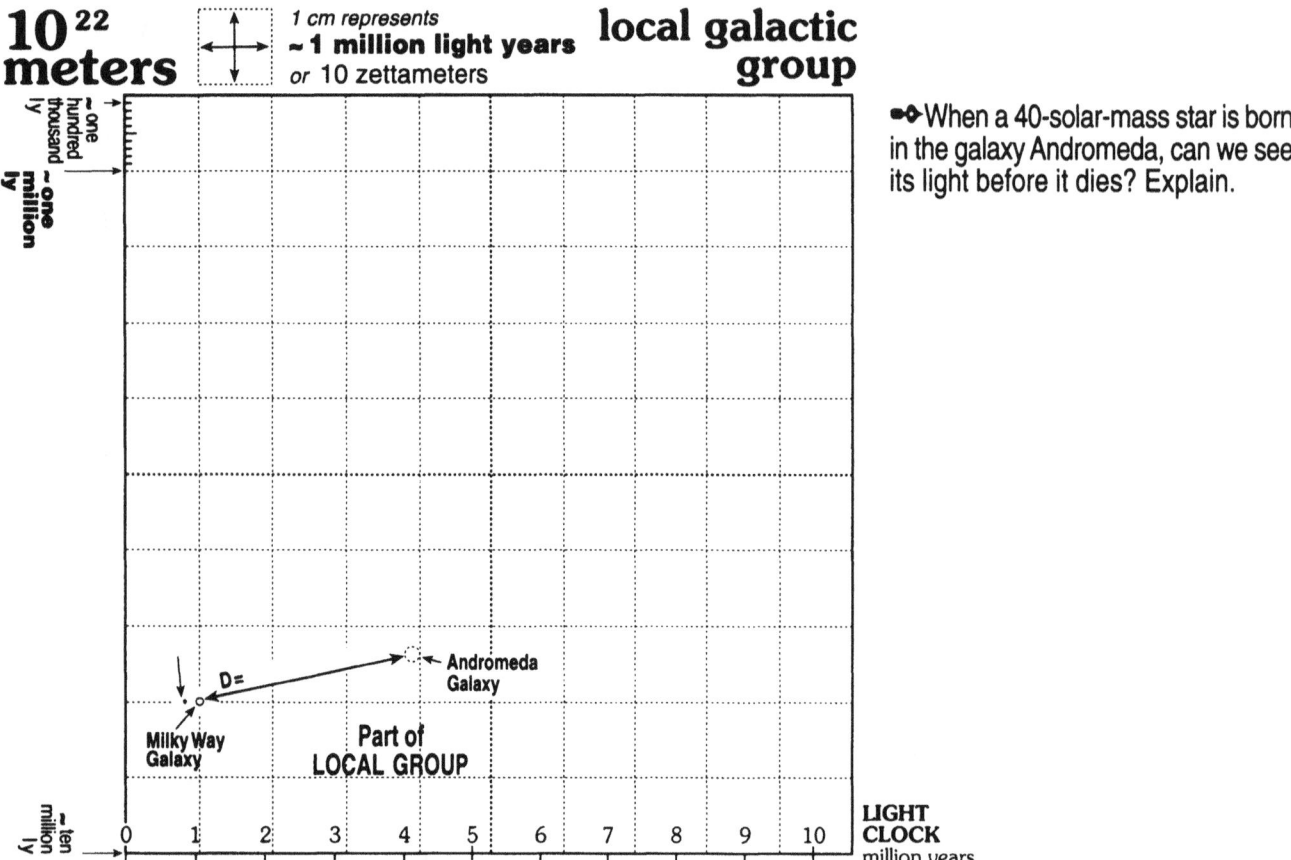

➥ When a 40-solar-mass star is born in the galaxy Andromeda, can we see its light before it dies? Explain.

10^{24} meters — galaxies like dust

1 cm represents ~100 million light years or 1 yottameter

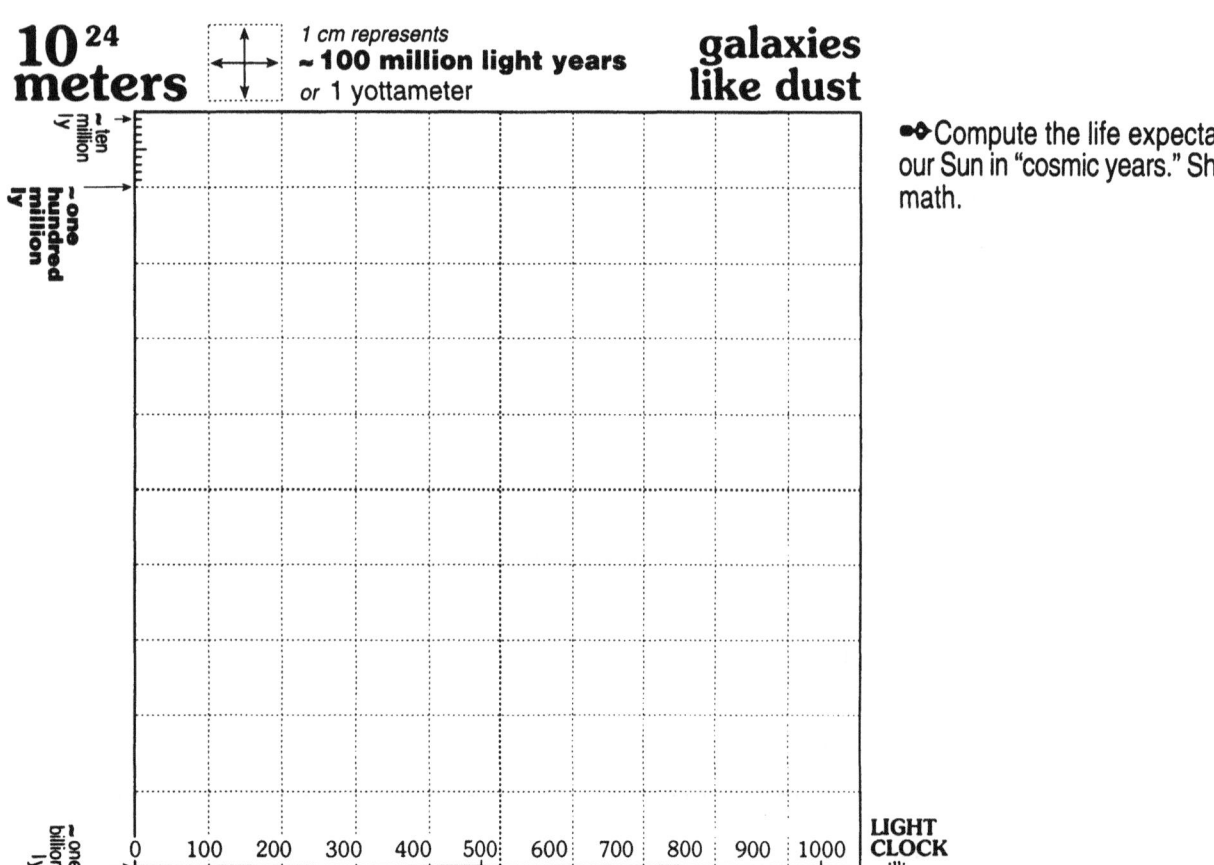

➥ Compute the life expectancy of our Sun in "cosmic years." Show your math.

↦ Draw GRB 990123 and AGN 3C 273 to scale @ 10^{26} meters (as you've already drawn them here.)

a. What do "GRB" and "AGN" stand for? (See distance tabs.)

b. Which is the short-term event? The long-term object? (See Time Tabs @ 10^{10} m and 10^6 m.)

c. Compare their energy outputs over time.

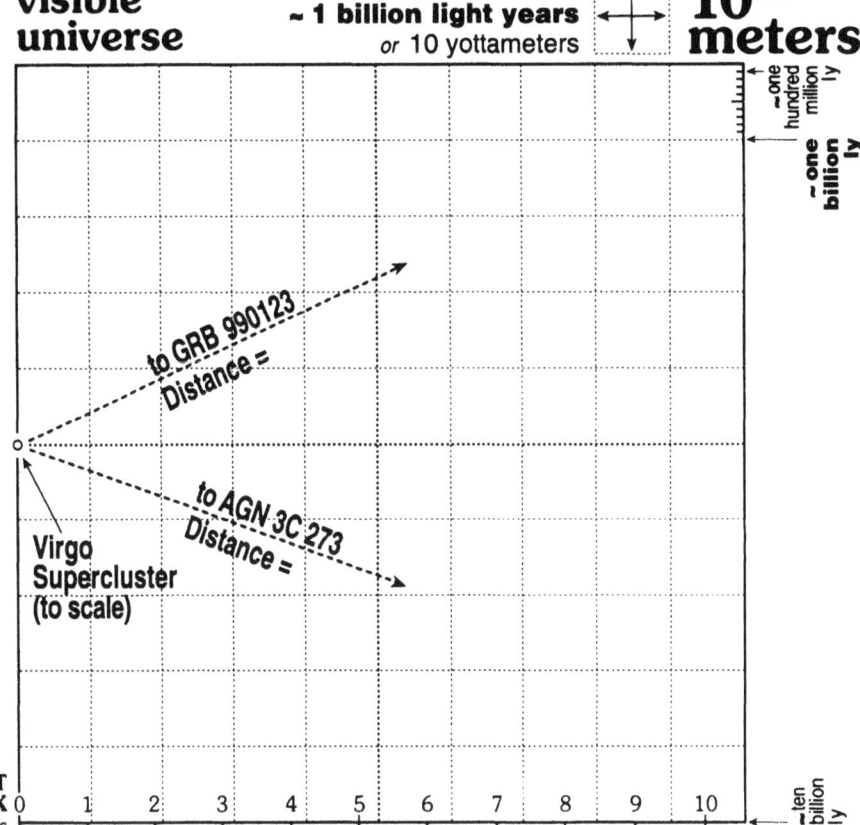

visible universe

1 cm represents ~ 1 billion light years or 10 yottameters

10^{25} meters

to GRB 990123
Distance =

to AGN 3C 273
Distance =

Virgo Supercluster (to scale)

LIGHT CLOCK   0   1   2   3   4   5   6   7   8   9   10
billion years

~ one hundred million ly
~ one billion ly
~ ten billion ly

---

inside BACK COVER

The back of this book investigates the
LARGEST OBJECTS!
LONGEST DISTANCES!
GREATEST TIME PERIODS!

Copy/collate 1 set of 12 double-sided pages per student.
This booklet is a component of the TOPS book SCALE THE UNIVERSE.
Copyright © 2005 by TOPS Learning Systems, Canby OR 97013

10^{26} meters

1 cm represents
~ 10 billion light years
or 100 yottameters

beyond imagination

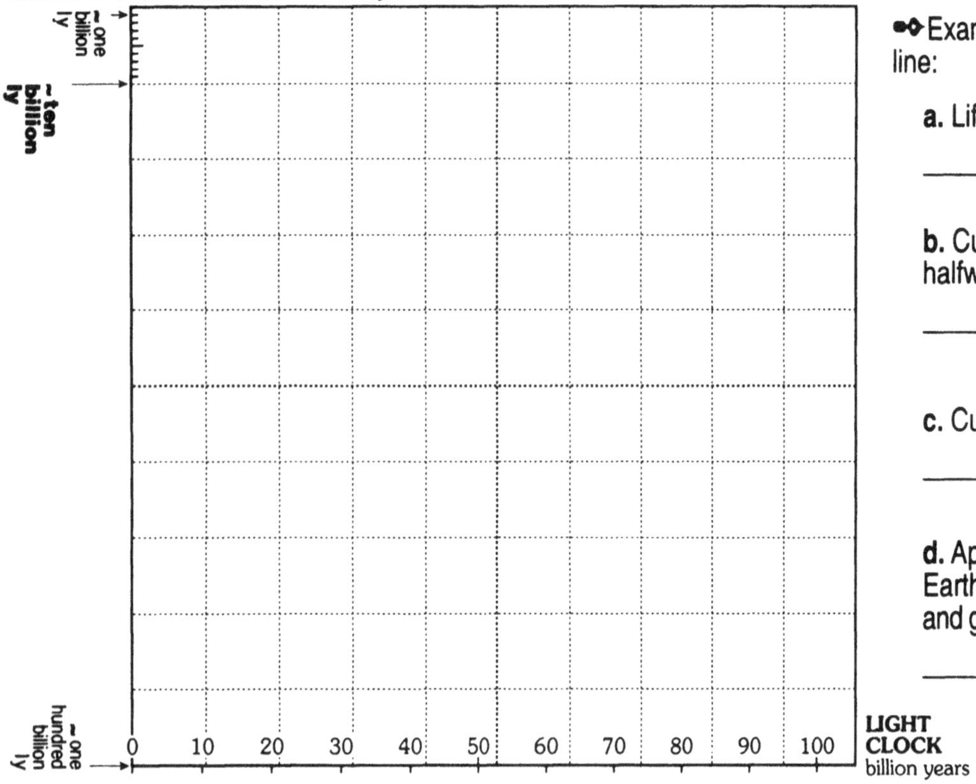

~one billion ly
~ten billion ly
~one hundred billion ly

0 10 20 30 40 50 60 70 80 90 100

LIGHT CLOCK
billion years

●◆ Examine Time Tabs to fill in each line:

a. Life span of the Sun?
_____

b. Current age of the Sun, now halfway through its life cycle?
_____

c. Current age of the Earth?
_____

d. Approximately how long it took Earth to coalesce from the dust and gas surrounding the early Sun?
_____

L: back

Activities C1-C5

BOOK OF SCALE

All Things Big *and* Small

LARGE to SMALL by powers of 10

name:

BACK

TIME TABS
("Sticky Tab" Version)

FAST: periods of electromagnetic radiation

PERIOD: time to travel one wavelength
UV RADIATION
100 times the energy of yellow light

PERIOD: time to travel one wavelength
Visible BLUE Light

PERIOD: time to travel one wavelength
IR RADIATION
1/100 the energy of yellow light

PERIOD: time to travel one wavelength
Highest Energy GAMMA RAYS
detected by GLAST's **LAT**
(Large Area Telescope)

PERIOD: time to travel one wavelength
Visible RED Light

PERIOD: time to travel one wavelength
AM "1000" Radio Wave
(1,000 KiloHertz = 10^6 cycles/second)

PERIOD: time to travel one wavelength
GAMMA RAYS
detected by *both* GLAST monitors:
(**GBM** and **LAT**)

PERIOD: time to travel one wavelength
MICROWAVE BACKGROUND RADIATION
left over from Big Bang:
1/1000 the energy of yellow light

PERIOD: time to travel one wavelength
Lowest Energy X-RAYS
detected by GLAST's **GBM**

PERIOD: time to travel one wavelength
FM "100" Radio Wave
(100 MegaHertz = 10^8 cycles/second)

PERIOD: time to travel one wavelength
Longest FCC Radio Wave
(no license required below 9,000 cycles/second)

CRAB PULSAR
galactic gamma ray source
spinning neutron star
(period = 33 ms)

A Long-Lasting GRB
(Gamma Ray Burst)
(photon shower lasts 1.5 minutes)

One EARTH ROTATION
on its axis

SUNLIGHT TRAVEL TIME
to reach Earth

Period GLAST ORBIT
(16 orbits per day)

One Human BREATH
average: 20 breaths/minute

GLAST Window
maximum time to observe a
single gamma source per orbit

One FORTNIGHT
(half of a moon cycle)

One EARTH ORBIT
around Sun

A Brief GRB
(Gamma Ray Burst)
photon shower lasts 3 ms

One Human HEARTBEAT
average: 72 beats/minute

One MOON CYCLE
full moon to full moon

AVERAGE: one year or less

15-Solar-Mass STAR
life span until nuclear fusion ends
(less than a Cosmic Year.)

HUMAN Average Life Span

Age of EARTH
(Greater than a cosmic year.)

Recorded HUMAN HISTORY

40-Solar-Mass STAR
life span until nuclear fusion ends
(The brighter they burn, the faster they die.)

GLAST Life Span
Gamma-ray Large Area Space Telescope
minimum expected useful life
(Best hope is 10 years.)

SUNSPOT CYCLE
solar storm activity:
maximum to maximum
(Less than 1 human generation.)

2/5-Solar-Mass STAR
life span until nuclear fusion ends

Precession of EARTH'S AXIS
one wobble of Earth's axis
(Polaris has not "moved" in recorded history.)

A "Year" on PLUTO
(about 12 human generations)

1-Solar-Mass SUN
life span until nuclear fusion ends

Age of Observable UNIVERSE

COSMIC YEAR
Sun revolves once around
Milky Way's center

SLOW: more than one year

HUMAN SCALE RULER — activity C4

		tiny skin pore	small freckle (1 mm)	finger-nail wide	hand width	1 giant step	10 giant steps	100 giant steps	12 minute walk	2 hour walk	2 day hike	20 day hike	200 day hike	5.5 year hike	55 year hike	10 life-times
The real you (*actual size*):		10^{-4} meters	10^{-3} meters	10^{-2} meters	10^{-1} meters	10^{0} meters	10^{1} meters	10^{2} meters	10^{3} meters	10^{4} meters	10^{5} meters	10^{6} meters	10^{7} meters	10^{8} meters	10^{9} meters	10^{10} meters
What if you were *larger* by 8 OM's:		10^{4} meters	10^{5} meters	10^{6} meters	10^{7} meters	10^{8} meters	10^{9} meters	10^{10} meters	10^{11} meters	10^{12} meters	10^{13} meters	10^{14} meters	10^{15} meters	10^{16} meters	10^{17} meters	10^{18} meters
What if you were *smaller* by 9 OM's:		10^{-13} meters	10^{-12} meters	10^{-11} meters	10^{-10} meters	10^{-9} meters	10^{-8} meters	10^{-7} meters	10^{-6} meters	10^{-5} meters	10^{-4} meters	10^{-3} meters	10^{-2} meters	10^{-1} meters	10^{-0} meters	10^{1} meters
Some other sizes; larger/smaller by ___ OM's:		meters	meters	meters	meters	meters	meters	meters	meters	meters	meters	meters	meters	meters	meters	meters

© 2004 TOPS Learning Systems

TIME TABS — activity C5
("Cutout Tab" Version)

FAST: periods of electromagnetic radiation (fastest)

- Period = $10^{-3.954}$ s = 111 μs — **Longest FCC Radio Wave** (no broadcasting license required below 9,000 waves/second)
- Period = $10^{-6.000}$ s = 1 μs — **AM "1000" Radio Wave** (1,000 KiloHertz = 10^6 cycles/second)
- Period = $10^{-8.000}$ s = 10 ns — **FM "100" Radio Wave** (100 MegaHertz = 10^8 cycles/second)
- Period = $10^{-11.204}$ s = 6.1 ps — **Microwave Background Radiation** left over from Big Bang
- Period = $10^{-12.699}$ s = 200 fs — **IR Radiation** 1/100 the energy of yellow light
- Period = $10^{-14.623}$ s = 2.4 fs — **Visible RED Light**
- Period = $10^{-14.819}$ s = 1.6 fs — **Visible BLUE Light**
- Period = $10^{-16.699}$ s = 20 as — **UV Radiation** 100 times the energy of yellow light
- Period = $10^{-18.398}$ s = 400 zs — **Lowest Energy X-RAYS** detected by GLAST Gamma-ray Burst Monitors
- Period = $10^{-21.699}$ s = 200 ys — **GAMMA RAYS** detected by *both* GLAST monitors: GBM and LAT
- Period = $10^{-25.875}$ s = 0.013 ys — **Highest Energy GAMMA RAYS** detected by GLAST's Large Area Telescope (LAT)

AVERAGE: one year or less

- Period = $10^{7.499}$ s = 1 yr — **One EARTH ORBIT** around Sun
- Period = $10^{6.406}$ s = 29 d — **One MOON CYCLE** full moon to full moon
- Duration = $10^{6.083}$ s = 14 d — **One FORTNIGHT** (half moon cycle)
- Period = $10^{4.937}$ s = 1 d — **One EARTH ROTATION** on its axis
- Period = $10^{3.732}$ s = 1.5 h — **One GLAST ORBIT** (~1.5 hours)
- Duration = $10^{3.380}$ s = 40 min — **GLAST Window** maximum time to observe a single gamma source per orbit
- Duration = $10^{2.699}$ s = 8.33 min — **SUNLIGHT TRAVEL TIME** to reach Earth
- Duration = $10^{1.954}$ s = 1.5 min — **A Long-Lasting GRB** (Gamma Ray Burst) photon shower lasts 1.5 minutes
- Period = $10^{0.477}$ s = 3 s — **One HUMAN BREATH** average 20 breaths/minute
- Period = $10^{-0.079}$ s = 830 ms — **One HUMAN HEARTBEAT** average 72 beats/minute
- Period = $10^{-1.477}$ s = 33 ms — **CRAB PULSAR** (Spinning Neutron Star: 30 spins/second) a galactic gamma source
- Duration = $10^{-2.523}$ s = 3 ms — **A Brief GRB** (Gamma Ray Burst) photon shower lasts 3 milliseconds

SLOW: more than one year (slowest)

- Life Span = $10^{18.277}$ s = 60 billion yr — **2/5-Solar-Mass STAR** nuclear fusion ends
- Current Age = $10^{17.636}$ s = 13.7 billion yr — **OBSERVABLE UNIVERSE**
- Life Span = $10^{17.540}$ s = 11 billion yr — **1-Solar-Mass SUN** nuclear fusion ends
- Current Age = $10^{17.153}$ s = 4.5 billion yr — **EARTH**
- Period = $10^{15.879}$ s = 240 million yr — **COSMIC YEAR** Sun revolves once around galactic center
- Life Span = $10^{14.499}$ s = 10 million yr — **15-Solar-Mass STAR** nuclear fusion ends
- Life Span = $10^{13.499}$ s = 1 million yr — **40-Solar-Mass STAR** nuclear fusion ends
- Period = $10^{11.915}$ s = 26,000 yr — **Precession of EARTH AXIS** (one "wobble")
- Duration = $10^{11.199}$ s = 5,000 yr — **Recorded HUMAN History**
- Period = $10^{9.879}$ s = 240 yr — **One "YEAR" on PLUTO**
- Life Expectancy = $10^{9.380}$ s = 76 yr — **HUMAN**
- Period = $10^{8.541}$ s = 11 yr — **One SUNSPOT CYCLE** solar activity: maximum to maximum
- Life Expectancy$_{(min)}$ = $10^{8.199}$ s = 5 yr — **GLAST Life Span** Gamma Ray Large Area Space Telescope (best hope is 10 years)

Feedback

If you enjoyed teaching TOPS please tell us so. Your praise motivates us to work hard. If you found an error or can suggest ways to improve this module, we need to hear about that too. Your criticism will help us improve our next new edition. Would you like information about our other publications? Ask us to send you our latest catalog free of charge.

For whatever reason, we'd love to hear from you. We include this self-mailer for your convenience.

Sincerely,

Ron & Peg

Ron and Peg Marson
author and illustrator

Your Message Here:

Module Title _____ Date _____

Name _____ School _____

Address _____

City _____ State _____ Zip _____

———————————————————————— FIRST FOLD ————————————————————————

———————————————————————— SECOND FOLD ————————————————————————

RETURN ADDRESS

_____

_____

PLACE
STAMP
HERE

TOPS Learning Systems
342 S Plumas St
Willows, CA 95988

TAPE HERE

www.ingramcontent.com/pod-product-compliance
Lightning Source LLC
Chambersburg PA
CBHW081940170426
43202CB00018B/2964